BUCKING
THE TREND

The Story of Buck Autrey and Miller Electric Company

BUCK AUTREY

WITH KENNETH R. OVERMAN

Dedicated to my greatest supporter in all my endeavors,
my best friend and wife for life, Betty White Autrey

Young Buck Autrey working in Jacksonville's old Independent Life building - 1953

Contents

List of Vignettes

List of "A Word About..."

Introduction

Have you ever wondered if the products or services touted in all those slick advertisements are really what they're cracked up to be? And are you like most consumers, who look behind the compelling façade and conclude *some* of what they promise *might* be true? While I don't believe all are guilty, the skeptic in me points to the track record; when a business makes an offer, we hope for the best and expect slightly less. After all, it's common practice to hide the flaws, right? Not necessarily. After I got to know Buck Autrey and Miller Electric Company, I realized they are an exception. Inside and out, they really are all they appear to be.

When I first sat down with Buck to record his story I was immediately drawn to his larger-than-life personality. The rumbling timbre of his southern accent, his broad, easy smile, and clear details of places and events long before I was born, convinced me I was in the presence of a true leader.

Plus, he delivered his anecdotes in flawless sequence ... a real joy for a biographer. We started with his earliest memories in Tallahassee and went on to his teenage job at a roller skating rink where he met Betty. "Marrying her," as Buck recalled, "was when my life really began." He went on to discuss the tumultuous years when the American electrical contracting industry came under fire by the US Department of Justice. After several months of recording sessions and a full box of AAA batteries for my digital recorder, Buck ended the story with his current role as chairman emeritus of Miller Electric Company.

Listening to Buck's experiences was the verbal equivalent of an "ethical will," a treasured document dating back to Genesis, when rabbis encouraged men to transfer values and life lessons from one generation to the next. After I interviewed several others in the company, I realized Buck had already established such a will. In fact he began doing so long before he joined Miller Electric. You'll find the evidence in stories told by family and colleagues later in the book.

Eleven others share their version of Buck's influence on their work and their lives. They are his sons, daughters, grandsons, associates, colleagues, and the CEO of the National Electrical Contractors Association. That they corroborated Buck's accounts came as no surprise after I got to know him better. At first their recollections struck me as over the top with praise and accolades. But as I interviewed one after another I heard essentially the same things, which is significant since truth can be found as much by what is *not* said as what is said. With so many witnesses to his life and legacy, I knew he stood head and shoulders above the rest.

Buck was a constant learner. As John Grau, CEO of NECA, pointed out, "He was a self-taught, self-made man and where he got his knowledge and wisdom is a mystery. When I talked with his son Ron or Miller's president David Long, they constantly quoted things Buck taught them about his philosophy of business or life." Many examples follow.

The man and his company went through some very hard times, which kept Buck on a constant learning curve. His studies went beyond the extremes, such as memorizing the entire electrical code book in bed late at night while Betty helped by firing questions at him.

When Buck wasn't learning, he was teaching. Employees said they learned a lot just by observing him. Again John pointed out: "They respect him for his sage wisdom, and so do I. They were the kind of principles found in books with quotes by famous people. It seems Buck internalized many of those philosophies, and I don't know if he heard them somewhere, read them, or came up with them on his own."

Sometimes Buck and I went to lunch after a morning interview, usually to Bono's Bar-B-Q in downtown Jacksonville, the original location established in 1949. Buck took his seat at the middle of the table with other long-timers including respected city managers, a former chief of police, politicians, and his lifelong associate, Ed Witt. I felt like an observer of more than sixty years of Jacksonville history while Buck officiated with the comfortable elegance of a patriarch. He also picked up the tab, which I'm told he always does.

This book is about legacy … the passing of a torch from one set of capable hands to the next. It is Buck's gift taught by his life at home, at work, around the campfire, and through this biography. It has been a real pleasure to record this dynamic story from the inside out, and I've never been so fascinated with the evolution of a strong character and hard-earned, well-deserved success. Literally hundreds out there agree with me and I believe you will too.

But now, let's hear his story.

Kenneth R. Overman

ONE
The Early Years

Jacksonville, Florida, 1966

"There wasn't a whole lot to celebrate because the company was dead-ass broke. In fact we owed a lot more money than we were worth, so it was kind of iffy in my mind whether we would make it at all. I was dedicated and I knew we were going to have to try. It was my best shot in life at that point … the only shot I would consider."

Those words reflect my exact thoughts that day in 1966 when I accepted my appointment as president of Miller Electric Company. By then, I had been with the company fifteen years and, in accordance with the process of succession, my appointment was a foregone conclusion. It would have been a big surprise to me if it didn't happen. Mr. Dandelake, our president, had been ill for some time and it looked as if he wouldn't recover. So I made a commitment to Mrs. Wynn, the daughter of company founder Henry Miller, that if she decided to buy Mr. Dandelake's shares of the company, I would commit to staying on as president— which she wanted me to do.

The transition itself was not a big deal. All we did was post an intracompany announcement to the employees—all forty of them—and issue a press release. Aside from that, I got a little plastic engraving of the announcement which we sent to our customers. I don't remember having any feelings of elation. There were no toasts of wine. There was no ceremony. My wife was of course happy as a lark to see it happen. Still, behind all that was a nagging question in my mind. Would we make it?

Thirty-Four Years Earlier

Back in 1932, Tallahassee, Florida, was just a small town. But on October 20 of that year I helped the population grow when I came into the world all red-faced and screaming. While my hardworking parents and their friends rejoiced at my arrival, there was little else to celebrate. America was deep in the Great Depression and someone like me coming onto the scene didn't have much to look forward to.

Thirteen years later World War II ended. I remember the day because it was all over the radio broadcasts and I saw it in the newsreels at the Saturday matinee. We went to the movies fairly often back when a quarter bought a ticket, a box of popcorn, and a Coca-Cola. There was always a double feature, usually a black-and-white cowboy movie, some drama movie, and a cartoon. They always showed the newsreel first, and that's where I saw what happened with the war and when President Roosevelt died and Harry Truman took over. Or the day Truman dropped an A-bomb on Japan, helping us win the war. I remember the photos of General MacArthur getting signatures from the surrendering Japanese on the USS *Missouri*. That made a huge impression on me.

I made it through the ninth grade, at which time my father pulled me aside and said, "Buck, that's all you need to learn. Now you need to go to work." That statement marked the end of my formal education and the beginning of a long job search. I banged around from one job to another until 1947 when, at the age of fourteen, I joined the Florida National Guard. Though I lied about my age, my mom was okay with it, thinking there would be no harm. My dad, on the other hand, was very upset when he found out, but I was already in and there wasn't much he could do. After all, it was a job.

Not long after I joined the Guard, a friend of mine told me about the advantages of joining the Air Force. I looked into it and decided to make the switch if I could. My friend who'd already joined helped me transfer out of the Guard and go directly into the Air Force. They sent me to Lackland AFB in San Antonio where three months later, when I was almost through with basic training, I developed pneumonia. It was a fairly serious case and I wound up in the hospital for eleven days. When my mother found out she got worried and contacted the military. She told them my real age, and that earned me an automatic discharge and a ticket back home.

My short military career was quite an experience for a young fella, and when I look back I realize I actually enjoyed myself. I learned a lot from the experience and believe it helped me grow up and learn some discipline.

I lived with my parents for a couple more years while I went from one job to another, doing anything I could to earn money. I had a paper route. I worked as an auto mechanic. I gained a lot of experience over those years but I also knew I could do a lot better. Although I was restless and wandered from job to job, I never had the problems many kids have today with alcohol, drugs, and that sort of thing. I did smoke for a while, but that was the only vice I had, if you could call it that.

Dating on Wheels

I was sixteen when 1949 rolled around, and by then I'd settled into a job with the Purina Feed Company, working as a salesman. I drove around town selling products in the daytime and spent most evenings at our local roller skating rink. I loved the sport and eventually landed a part-time job as a rink floor manager. One of my duties was to help kids put their skates on—the old steel roller skates with fastenings that clamped onto the leather sole of a shoe.

One night I was down on my knees fitting skates for some overweight kid, when I glanced up and saw a couple of great-looking legs. They were connected to a red miniskirt above and white skating boots below. I followed the image as she skated past and I nearly crushed the kid's toes from tightening the wrench too hard. When he was squared away, I quickly put on my skates and rolled out to the floor to find those legs.

Back then, the floor manager was kind of important to most skaters. I wore formal clothes and cruised around on nice boot skates, and with that kind of profile I sidled right up next to her and started talking. We just skated around and around, chatting and having a good time. I was impressed. She was pretty good on skates and I liked her name … Betty.

The evening flew by. We skated until the rink was about to close and we had to say good night. A couple of days later she showed up again, and again we spent quite a bit of time talking. She came night after night and we skated together, talked, and sometimes grabbed a Coke at the snack bar. When we were more comfortable with each other we started to experiment with a few dance skating routines. We had a lot of fun and really seemed to hit it off.

On our first date we drove to the movies. I didn't have an automobile, but I did have a truck from my job at the Purina Food Company; so we decided to drive her little English Ford. After the movie we parked in a lovely area overlooking Lake Ella on North Monroe Street. We watched the ducks waddle around and paddle into the water. We just sat, talked, and spooned (spooning meant hugging and kissing a little bit). It was probably the first time we kissed. Yeah, it was great!

Like most young couples, our conversations revolved around ourselves: what we liked, what we wanted in life, school, sports, skating, and our families. She said she was the oldest of three siblings, with a younger sister and a brother. She was fairly mature for her age, actually, and bragged about getting her driver's license when she was twelve or thirteen. I got mine when I was fourteen, so obviously both of us lied since the legal age was sixteen.

Betty told me her father was a superintendent with a company called Miller Electric and had worked many years for them. She said he was one of the first employees the original owner, Henry Miller, hired after he opened the company in 1928. Her father worked on several jobs during his career with Miller, mostly in Jacksonville. At the time we were dating, he performed electrical work at the new Florida A&M College Hospital, which impressed me a lot.

While Betty and I skated and dated, we made a lot of friends at the rink. Eventually a group of guys and gals got together and decided to put on a show for the March of Dimes. We had a couple of friends from the local university help choreograph the program, and wound up with a little skit we performed for charity. For our part, Betty and I did a kind of country-western number complete with costumes. We also added a variety of Broadway tunes, like "Murder on Fifth Avenue" and "South Pacific" … stuff like that. The show turned out to be a country-western variety show with about a half-dozen themes … all on roller skates.

The first show in the fall of 1949 was as successful as you could expect for a small town like Tallahassee. After all, it was just a bunch of kids getting together to do something helpful, and we had a lot of fun doing it. We did a couple of shows

Seventeen - where it all began

The Skating Road Tour

in our town and later took it on the road to Birmingham, Alabama, where we performed at the Lowe's Theater and Skating Rink. That too was fairly successful … not bad for a bunch of kids having a good time.

Aside from a few friends at the rink, Betty and I were basically loners. She never had many friends and I didn't have many friends other than my skating buddies. We never went out with other couples, so we dated by ourselves—unless her mother came with us. Yep, it was like the old song; whenever we were on a date (except for that first one at the lake), her mother was almost always in the backseat. Her parents were very protective of her, and I don't know if it was because they didn't trust her, me, or both of us. That's how it was.

The Deal Is Done

I proposed to Betty one night at the drive-in movie, sometime near the end of *She Wore a Yellow Ribbon*, starring John Wayne. Betty always said I didn't really propose to her. She insisted I only brought it up in conversation: "When we get married we're going to do such and such," to which she replied, "When we get what?" No matter who said what, we finally decided to get married.

My next step was to confront her father. Of course he wasn't too happy about it, to say the least. His reply came as no surprise.

"My daughter ain't marrying no skate rink jockey!"

"Mr. White," I said, "I'll make a deal with you. You work out some way I can get a job and I'll give up the skating rink. But we are going to get married and I want your blessing."

He never asked if we loved each other. He only said, "Well, I guess y'all understand each other."

"Yeah," I said. "We understand each other pretty well and we know exactly what we want."

"Well, okay," he shrugged.

None of our folks were happy about it, really, because all of them were concerned about our age. We were both seventeen. She turned seventeen on May 16 of 1950 and I had turned seventeen on October 20 of the previous year. It still didn't matter to us because the deal was done.

The original plan was to get married on skates, but that changed. Betty attended a Baptist church on the north side of Tallahassee and her girlfriend was also a member of that church. I think the two of them decided skates weren't the best way to honor her marriage. So in the last moment "we" changed our minds.

We were married in her small, white Baptist church on June 9, 1950, at 6:00 PM … about two hours before Betty graduated from high school. Anne Draughn, Betty's good high school friend, was her bridesmaid. Al Skinner, a fella who worked with me at the Purina Company, was my best man. We didn't have a lot of money, but since we planned to move to Jacksonville right after the wedding, I'd already resigned from my job at the feed company. After the reception we hopped into Betty's car and drove twenty miles to a motel in Quincy, Florida, and spent one night. That was our honeymoon.

As for keepsakes from the wedding, her parents didn't want us to take photos because, as her father said, "Y'all's marriage will be over before the pictures are developed, anyway." The only wedding gift we received was an electric iron my employer's partner gave us from his store in Quincy. So that was all there was to it. One year after I first saw those gorgeous legs in white roller skate boots, we were married.

I couldn't know it at the time, but I later realized my life actually began when I married Betty. From that day on, all kinds of opportunities opened up for me. It would be quite a while before her skeptic father would come around, but it finally happened on our fiftieth wedding anniversary, when he pulled me aside.

"You know Buck," he said, "you turned out to be a fairly good son-in-law after all."

It took him fifty years to say that, but I really appreciated it.

Lineman Buck

We lived with Betty's folks in Jacksonville until August, when we rented our own apartment. The week after we married I started working at any odd job I could find. During the week I pumped gas at a filling station and on weekends I worked as a part-time butcher at Setzer's meat market. A friend who worked at

Our first wedding anniversary

First anniversary - 1951

Setzer's also had a job at the traffic signal division in Jacksonville. He helped me get a summer job there through some kind of city commissioners' help program.

After two weeks in the signal division's line construction department, I decided to try for a job where Betty's father worked. I went to the Miller Electric shop, found the warehouseman, and told him who I was and what I wanted. He made a call upstairs and told his boss that Mr. White's son-in-law worked for the city's line construction department and wanted to work for Miller.

"Great!" said the boss. "We need linemen."

They gave me a referral to the union so I could work for the company.

That was August of 1950, and my first position with Miller Electric was as an apprentice groundman—a "line grunt," as they called it. As a union employee I started out at 85 cents per hour on a five-day, forty-hour week.

My first job was on Philips Highway, working as a groundman servicing lines on utility poles. I was on that job a few weeks, when a guy I worked for—a lineman named Slim Stevens—was electrocuted. He was a young fella in his late twenties, with a wife and a couple of kids. I had just finished tying off what they call a down guy, a wire supporting the pole to the ground. Slim was up top and somehow came in contact with both the high-voltage line and the down guy at the same time. The instantaneous grounding sent several thousand volts stampeding through Slim's body, killing him instantly. The way things were on that particular job, had it happened five minutes earlier, *both* of us would be dead.

The other guys on the job and I performed respiration attempts on Slim until the ambulance crew took over. They worked on him a while, and kept on working while they put him on a gurney and into the ambulance. I climbed into the back and we took off. As we sped to St. Luke's Hospital on the Northside they kept trying to resuscitate him, but Slim never took another breath.

The accident happened in the late afternoon, right before my wife was due to pick me up, and she arrived just as the ambulance pulled away. Betty asked the foreman where I was and he told her I was in the ambulance. She had no idea if I was injured or not ... only that something bad happened. She took off behind the ambulance, followed it all the way to St. Luke's, and pulled up behind it at the emergency entrance. When I stepped out she was of course upset ... which in turn upset me even more. To see my wife like that and realize how close to danger we were in that profession really bothered me. Minutes before, Slim was alive and talking as if there wasn't a care in the world. Now he was dead. It was very traumatic.

After that incident Betty and I decided line work was not my forte, so I petitioned the local union to allow me to work the inside wireman division. They said I could, as long as I continued to go to school and keep my union card valid for outside line construction jobs … in case they needed a registered apprentice (regulations required registered apprentices to be available for a proportion of certain government jobs). I readily agreed to both the inside and outside units. Looking back, I realize it presented a big opportunity for me since I could go to school four nights a week and keep working. That was how I learned both journeyman wireman and journeyman lineman skills at the same time.

Meeting Mr. Miller

Sometime after Slim Stevens died, I was assigned to work on a line gang on Philips Highway. The old power lines along that route had to be dismantled to make room for new lines. Our job was to remove the wires, the crossarms, and all else from several hundred old forty-foot poles, and give them back to the city. One day our winch truck operator fell ill, and the foreman assigned me to drive it in his absence. By then I had some experience working on the line for the city and knew how to drive trucks, so it was an easy switch. He said the idea was to dig all around the pole, hook the winch truck to it, pull it out of the ground, and lay it down. Another worker was assigned to me and we got to work.

It was a hard job and took at least an hour to get each pole yanked out and placed on the ground. The going was so slow that I thought there had to be a better way. I kept working until I came up with a laborsaving idea, and then told my helper what we'd do. We dug a single hole behind the pole and used the truck to push it back into the vacant space. Next I hooked the line on the pole from the rear of the truck and put pressure on the winch by raising the front of the truck several feet off the ground. The last step was to literally bounce the truck up and down several times until the pole popped out of the ground.

Roy White's union dues receipt

The first pole came out in fifteen minutes. After that, we pulled from three to four poles every hour … a real laborsaving system. From a distance we must have been a sight—with our truck rearing on its back wheels like a stallion.

Along came Friday afternoon, when Mr. Miller and Mr. Dandelake visited the jobsites to pay the men off. Checks weren't allowed by the union, so we were paid in cash wherever we were on the job. Right when I had the front of my truck about four feet in the air, bouncing up and down, a green Cadillac pulled up right in front of us and two men climbed out. Although I knew Mr. Dandelake, I'd never met Mr. Miller until that day. But I knew who he was. They brusquely walked toward my truck while Mr. Miller frantically waved his hands in the air. I eased off and lowered the truck to the ground, wondering what was up.

"Young man!" said Mr. Miller in his deep southern accent. "What are you doing to my truck?"

"Mr. Miller," I said, "I'm pulling the poles out of the ground the way we're supposed to, by digging a hole and pulling it out and—"

"Well, yeah," he interrupted, "but I don't want you to tear up my truck. You keep that truck on the ground, you hear?" They turned around and left without another word.

About an hour later the same Cadillac pulled up. This time Mr. Miller walked over by himself. "Young man," he said, "you go back to working those poles just like before. You're doing a good job. … I didn't realize you were pulling so many per day."

I don't know if he knew I was Roy White's son-in-law or not. If not, he probably found out later. Had he known, he might have approached the situation differently since Roy was good friends with Mr. Miller. I found out later the company received a fair amount of money for each pole they turned in to the city. The more poles, the more money the company made. By the time we reached our stride using my method, we pulled between twenty and twenty-five poles every day.

My first meeting with Mr. Miller could have been under better circumstances, but it worked out quite well anyway.

Apprenticeship and Onward

Up until I married, I was never satisfied with my home life or my jobs. However, as I said before, when I married Betty my life really began. It seemed marriage helped me line up my priorities and at the same time opened doors to many possibilities. First, I was able to go to school, and that's where I discovered I was a pretty good student. Everything I learned made me want to learn that much more, and faster. I was good at mathematics. I understood the basics of electricity. I thrived in the apprenticeship training I received through Miller Electric. The more I learned, the more interested I became.

I attended school four nights a week and, in so doing, accelerated my training. I graduated in three years, and in 1954—four years after joining the company—I received the Outstanding Apprentice of the Year Award for Best Academic Performer. That was also significant because the trade school was sponsored by both Miller Electric Company and the International Brotherhood of Electrical Workers union, or IBEW. Since all instructors were members, they worked their trade during the day and taught school at night. Likewise, apprentices worked their trade during the day and attended school at night. With such close scrutiny, the students' good merits—or the lack thereof—were apparent.

I didn't stop my education after graduating from apprentice training. I enrolled in a correspondence course in electrical engineering through ICS, the International Correspondence School. I studied two divisions of basic electrical engineering and powerhouse generation, among other things. My third learning system involved the transmission of electrical currents, transmission lines, and electric systems construction. When I completed all that I kind of lost interest in going further. I had already progressed to the level of estimator and now worked in the office.

I recall our early years when Betty and I were awake in bed and I kept the light on to study for my exam. She stayed awake with me, asking questions from the codebook for me to answer and memorize. There were many such times and I believe she actually helped me memorize the entire National Electrical Code from cover to cover, nearly word for word. Those sessions helped me pass every one of the electrical exams. I never failed one, and always passed by a large margin.

Need I say more? Betty is a great woman and she has been my driving force in the development of my career and my life.

Growing Deeper

I worked for my father-in-law on the Independent Life Building (presently called the Wells Fargo Center), a skyscraper in downtown Jacksonville, until I became a journeyman in 1954. Sometime after that I was transferred to Kings Bay, Georgia. The purpose of the job was to prepare the facility to receive hydrogen bombs from Aiken, South Carolina, to be loaded onto ships or submarines bound for parts unknown. That never happened because they changed the location of the project and they stored ammunition in any of the bunkers strewn throughout the area. When the project became dormant, they switched me over to the St. Regis paper mill as a foreman. While there I was promoted to assistant superintendent in charge of around 100 people. I stayed there a year, and returned to the Jacksonville office to work on a service truck as a journeyman electrician. I did that for another year, until I was transferred to the service department.

Apparently Mr. Davis, the manager of the service department, thought I showed talent in pricing my own jobs. The result was in 1956 I started to price jobs for him while making my service calls at the same time. It wasn't long before I pretty much ran the service department as well as keeping up with my service truck work.

Not long after I started working inside, company president Jim Dandelake also thought I had some talent and appointed me assistant under a senior estimator named Ken Stowe. Although I had to keep up with my main service truck job, I learned a lot from Ken because he was brilliant at his job. He developed a powerful method of estimating that stuck with me throughout my career.

One day in the early stage of my apprentice estimating, Ken took me to lunch at the Green Derby restaurant in Jacksonville. He said he wanted to continue talking about the procedures of estimating and fill me in on what he expected of me. The restaurant was full, so we sat at the bar and ordered our food. A large TV on the opposite wall carried the October 1956 World Series between the New York Yankees and the Brooklyn Dodgers. In those days the teams played baseball

Line gang winch truck for pulling poles

Jim Dandelake

in the daytime, so people had to grab parts of the game wherever they could. We got our sandwiches, ate, and when we were done I was ready to go. I looked over at Ken, who continued to watch the game, so I cleared my throat and said, "We need to get to work, don't we?"

"No, not yet," he said, with his eyes glued to the screen.

Around the sixth inning Ken pointed and said, "He's pitching a perfect game!"

Lo and behold, we ended up watching the only perfectly pitched game in the history of the World Series. The Yankee pitcher was Don Larsen, and they won the series 4-3 over the Dodgers.

I'll never forget it. Nor will I forget how I felt for spending so much time away from work. I was convinced we'd get fired for staying there all afternoon watching a ball game. Still, Ken wasn't worried a bit. We returned to the office and stayed on the job until about eight or nine that night.

"That's the way we do business here," said Ken as we wrapped up the day. "We work until we get the job done, no matter how long it takes."

That day turned out to be my initiation into the company culture, and I enjoyed working with Ken in the few months I was with him. He was a good instructor whose example helped me realize hard work and loyalty to the company would be my number one priority.

It Comes with the Job

As my work and training progressed I gained experience on a variety of jobs. In fact, there's hardly a major building in downtown Jacksonville that doesn't have the Miller Electric footprint one way or another. Either we built it, wired it originally, rewired it, or pulled maintenance in it. Whatever the case, I was fortunate to work on several of those earlier projects. Some of the memories aren't too sweet, though, like the job at the Independent Life Building on the corner of Church Street and Julia—my first high-rise project. It was a seventeen-story building, and I was an apprentice working with a guy named Johnny Lesso, a journeyman welder. I remember too clearly watching one of the laborers get a heavy, sharp steelworker's spike clean through his skull. Killed him instantly.

Later on, Johnny and I were up on the third story of the same building, tacking down electrical ducts for raceways. We heard a loud crash right behind us. I turned around in time to see a huge 12" x 12" timber that apparently came loose from several stories above, as it fell through our floor and down to the next. It landed on a large table with several carpenters working nearby. I was amazed no one was injured … if we were six feet closer it would have killed us for sure.

We also had a near miss at the Winn-Dixie warehouse. As a journeyman I was assigned to work in the "banana room"—a large refrigerated area for curing South American bananas before delivery to retail outlets. They loaded the room with green bananas, sealed the cracks and airways, and burned wood charcoal in a stove in the center of the room. The carbon monoxide fumes from the coals quickly ripened the bananas.

One day my helper and I worked on a faulty circulating fan and had to stay inside the room several hours. Halfway through the job I looked across and saw my helper keel over and fall to the floor. He'd passed out cold. The moment it took me to realize what happened, I started to feel woozy myself. I came down off the ladder fast and noticed the heavy door was closed tight. Fortunately it was a refrigerator-type door, so I pushed my hip against the safety handle and it popped right open. Although on the verge of passing out, I grabbed my helper, dragged him outside to fresh air, and leaned him against the wall.

As carbon monoxide goes, if we hadn't cleared out of that room within five minutes, we both would have been gone. No one would have known until several hours later … if someone drove in with a forklift to discover dead meat on the floor alongside the bananas.

***

"Gramps" Norton was a foreman and a good friend of the family, so we had the kind of relationship that allowed us to have fun on the job now and then. Once during my journeyman days after I completed a job with Gramps, the company owed me $400 in overtime pay. Gramps, who loved to play jokes, decided to pay me off with one-dollar bills he'd stuffed in a paper sack. They were all jumbled up and crumpled when he proceeded to dump the entire sack of ones on the floor in front of me. Rather than quickly gather it up, I sat down on the floor and slowly unfolded each one, smoothing them out all nice, neat, and flat. After a while Gramps got real restless because I took about forty-five minutes of his work time. He eventually laughed it off.

The Day I Was Fired

The one who would become a strong business partner and friend, Ed Witt, joined Miller Electric Company in 1957. We worked together on the trouble truck and one of our jobs was to wire Mr. Dandelake's brother's office building on Atlantic Boulevard in Jacksonville. His brother was the controller for the city, and John Payton, my supervisor, said my job was to be sure whatever he wanted got done. He also told me to keep good records of our work.

One day when we were almost finished, John drove up to the site and told me the job was over labor costs by almost 100 percent. He also said Mr. Dandelake was very upset and he had to lay us off.

"Really?" I said. "Did you tell them what you told me to do … that everything his brother wanted done, we should be sure to do?"

"Yeah," he said, "but I don't have change orders or anything."

I walked over to the truck and pulled out a folder.

"You haven't been to the job in a while," I said, "so here's the book of change orders signed off by Mr. George Dandelake. I priced them out as best I could, and intended to give them to you at the end of the job."

He shook his head. "I don't know … it's probably too late now anyway."

I turned to Ed. "You hold on for a bit. I'll be back."

I jumped in my truck, drove to the company, and walked straight into Jim Dandelake's office. I told him his brother wanted the extra work done at his office as well as his home, and that I'd kept a record of the changes. I said there were several thousand hours of work that were far beyond the contract, and I kept accurate records of that too. I handed him the folder and after he leafed through it, he immediately called his brother. George agreed with what I wrote in the folder and said he even kept copies on his desk. He told Mr. Dandelake what a great job I'd done—both Ed and I—with everything he needed. He even complimented Mr. Dandelake on selecting such a fine young journeyman to do his work. Needless to say, Mr. Dandelake rescinded his order to fire us.

Wiring the Military Base

We did a job down in Green Cove Springs, a signal job at a military base, installing a monitoring and fire alarm system in a hangar. Ed Witt and I worked that job along with Bob Kegebein, another journeyman. We placed racks up on the poles. I climbed about thirty feet wearing only inadequate low-heeled shoes. After climbing fifteen to twenty poles, my feet were almost raw. Well, it made sense to me to keep an employee in good shape so he'd keep working. After work I drove over to the Birdseye Boot Company on Broad Street in Jacksonville and bought a pair of climbing boots. I told them to send the bill to Miller Electric Company, which was a common procedure at the time.

By the end of the job we had saved the company upward of 75 percent for labor because we worked hard and were motivated to get it done quickly. But Jim Dandelake told my supervisor I had charged a set of boots to the company and he didn't intend to pay the bill. John explained to Dandelake that we actually saved him about $20,000 worth of labor on the job from working so fast. He also told him he should go ahead and approve the boots. Dandelake did, and even complimented us on a good job.

The Fifty-Eight-Year Apprentice: Ed Witt Sr.

I got married in 1957, after completing a four-year hitch in the US Air Force. A month after the wedding I landed a job at Miller Electric, where I got to know Buck Autrey. Today, after fifty-eight years—more than a lifetime for many people—I can say it's been a very enjoyable and rewarding experience to be associated with Buck.

When I came on board, Buck was a superintendent at the St. Regis Paper Company in Jacksonville. When the project was completed, he returned to work in one of our service trucks where, as an apprentice, I was assigned to work with him. Buck also taught apprenticeship at night, so along with spending all day together I got to listen to his instruction two nights a week. Needless to say, we got to know each other well and over time found we had a lot in common. We both liked to work hard, and we both liked to hunt and fish. Later on we lived side by side in our lakeside homes in Keystone Heights, where we fished on the weekends, relaxed, and recharged our batteries for work on Monday.

During the '50s and '60s there were few employees at Miller, so we worked a lot of hours day and night, often on Saturdays. You might say we spent more time with each other than we did with our wives. That was the truth … we'd often get home around 11:00 PM, only to head out to work at seven the next morning and do it all over again. Sometimes it was difficult, but that's what made it fun and rewarding. When I climbed into bed at night I knew I'd accomplished something.

Although Buck and I did a lot of things together, we didn't seem to get in each other's way. Betty—Buck's wife—and my wife, Millie, also saw a lot of each other. His family and my family worked during the day, and in the evenings and weekends we all played together.

When Buck left the service truck to work as a project manager and estimator, I was assigned to help Bill Gothard, a journeyman. When Buck landed a job, he made sure Bill and I did the work for him. While that was a good arrangement, it wasn't always easy. Buck often bid too few hours and then challenged us to see how quickly we could finish. Sometimes he didn't give near enough hours to get the job done, but he'd only say, "I know you guys can do it."

That's how it was for a couple of years. Buck bid the jobs, estimated them, and we worked the projects. It turned out to be a real help because I picked up on Buck's estimating and bidding methods.

I had a good break in 1966 when I worked the Federal Office Building in Jacksonville. Buck was the project manager, and when the job was almost complete he asked me if I wanted to come into the office to work as an estimator. Of course I was happy he asked. I thought about it a couple of days, and finally told him since he'd always done right for me I decided to accewpt his offer. After that, I worked closely with

Buck and others on the management and estimating side of the business. What followed was a period of interesting change and growth for Miller Electric. Mrs. Wynn—Mr. Miller's daughter—handled the bonding, insurance, and finances, while I did the estimating. Buck, meanwhile, took care of sales, among other things.

People had confidence in Buck, and if he told them he would do something, he always did it. That kind of integrity went a long way to regain the confidence of some of our previous customers.

Miller experienced many personnel changes over the years. People grew older, retired, or passed away, and to replace them we watched the apprenticeship classes, where we picked the cream of the crop. Once we were confident with their capability, we brought them into the office and made project managers or estimators out of them. Next we gave them an opportunity in the same way Buck gave me. Buck definitely had a talent for picking good people, and through the years he found some excellent ones … which is obvious, since he picked me!

Many companies in our industry require their management to be qualified engineers before they even look at them, but our approach has always been different. Most of our management employees came inside from the field after we observed them during their apprenticeship and work performance. We saw how they worked, studied, and how interested they were in their job. The system paid off and our company is now one of the few providing on-the-job training. Many of our people moved up through the ranks to the level of vice president or president. One example is David Long, who started in the apprenticeship program and today is the president of Miller Electric. Buck came up through the ranks in the same way.

Miller Electric has been deeply involved with the National Electrical Contractors Association (NECA) since the beginning, and Buck got the company even more involved during his leadership. He believed if we could teach the competition through our association with NECA, at least we would know where they were at and what they were doing.

After Buck taught apprentice school, he worked his way through the ranks of NECA to become a chapter governor, chapter president, and later, vice president of District Three for NECA National. He ultimately became president of the association, representing the whole country, and served in that capacity for six years.

As Buck progressed through NECA, I kind of followed along. I was on the National Management Development Committee for NECA, served two years on the Council for Industrial Relations, and became contractor representative on the Joint Apprenticeship Committee for twelve years. Later I served on the Negotiations Committee for labor agreements in various jurisdictions. Like Buck, I also became governor of the NECA North Florida Chapter for sixteen years, and then vice president–at-large for NECA.

In the turbulent antitrust years Buck was away more often than not, so I stayed to help run the company. I communicated with him every day by phone wherever he was and, in so doing, helped him maintain company involvement.

Buck and I often discussed jobs and the challenges that came with them. If one of us didn't think we should be involved in a project because it looked too complex for our people to manage, I'd ask around for guys willing to pack up and move where we needed them. They would often say, "Yeah, we'll go where you go." They'd move with their families to wherever we needed them, and in some cases they returned to Jacksonville when the job was done.

Our policy was, "No matter where we worked, we always tried to take our own people first." Beyond that, if there were good local workers we could hire, we would. We'd simply go to a local area union, pick the best of the best, and train them along the way. Eventually they became employees we could depend on to go anywhere we needed.

Sometimes we stayed on the same job twenty or thirty years because they were happy with our work. For example, we had a job in Arizona that started in 1981 and lasted through 2000. That was one of many such jobs that started with a good relationship years before, and which also goes back to our relationship with NECA. Due to Buck's presidency and our relationship with so many in the industry, no matter where we were, the local NECA members trusted us. We served on their Union and Apprenticeship Committees, and were involved in productive negotiations through the years. So much of our success in other NECA areas came through Buck's leadership, by going outside of our region to engage with the rest of the industry.

So it all goes back to Buck, really. Fifty-eight years is an awful long time for two people to work together, yet we've had a strong run and a great relationship. That's why I tell people I was an apprentice in 1957 when I met Buck, and I've continued as his apprentice for fifty-eight years.

Ed Witt Sr.

Chairman, Board of Directors

A Word about Honesty

I believe honesty is number one. And right up there with honesty is integrity—walking your talk, so to speak. Honesty is the first attribute I look for in anyone, and I can tell during an interview whether or not they're an honest person. I've always had that ability and have made very few mistakes in that regard. I believe it's important because a liar and a thief—the opposite of an honest person—can destroy a company.

Right away, if someone is not honest, I don't deal with him. I don't care how good they are or how great they think they are. I won't have anything to do with them. Again, honesty is the basic foundational attribute and the first thing I look for in a person. Of course there's a difference between honesty and making mistakes. People can make mistakes along the way, but if they do, they should correct them real quick.

We stress the need for honesty and integrity in all our discussions, since the reputation of Miller Electric Company is at stake with everything we do. For example, if someone falsifies any document whatsoever, they place the company's reputation on the line. We've had some of that in the past and we've had to correct

Ed Witt at new office, 1991

it. Sometimes people do things out of desperation, thinking, *It's okay, just this one time.* No, it's never okay. If someone does a bad job or makes an honest mistake, the best thing to do is communicate it as quickly as possible to their peers or supervisor so they can help. The sooner it's taken care of, the less chance the problem will turn into a monstrous situation down the road.

Honesty as a value for good business comes from a long history of experience, forged through many years of trial in the workplace. A few examples will follow.

TWO
The Traumatic Years

Roots

The founder of the company, Henry Miller, apparently had some strong political connections. For one, he negotiated with Sen. Spessard Holland, a four-term Florida senator who helped Mr. Miller gain many contracts with the government during the war. The result was contracts at the Gibbs Shipyard, the Merrill Stephens yard, the Savannah Shipbuilding and Drydock Company, the Charlestown Navy Yard, and the Norfolk Naval Shipyard. We even installed electrical systems in the famous Liberty ships for the St. Johns River Shipbuilding Company in Jacksonville. That yard produced 82 of the country's 2,700 Liberty ships, which translated into a lot of work for Miller. Aside from continuous employment, our company also earned a Presidential Award for wartime productivity.

During those years Miller Electric established our own manufacturing plant called the Jacksonville Metal and Plastic Company. Because of the war, some materials were so scarce the company couldn't get switchboards from the manufacturers, so the creative guys at Miller built them from scratch. We designed them, built them, and wired them up. They were successful during the war, and that company grew to become a strong organization for a good period of time after the war ended.

The *Caribe Queen*

One of the toughest things to hit us in 1956 was the *Caribe Queen* job. The *Caribe Queen* was an old 400-foot freighter we were tasked to rewire to bring it up to Coast Guard standards. The wiring part was fine, but the problem was the contract Mr. Dandelake negotiated with the owners of the *Caribe Queen*. The ship fell under the Coast Guard maritime laws, which meant he was unable to secure a lien on the ship before it left port.

After we finished the job, the *Caribe Queen* slipped out of the downtown Merrill-Stevens shipyards and proceeded to conduct its sea trials. It never returned, but instead cruised on up to Baltimore. They never paid us our money, but the shipyards got their pay before it left the dock. The problem was we didn't have a contract with the shipyards; we had a contract with the owners, which was a mistake. We fought it in court, but I don't know what the final judgment was. Although I was a wireman at the time, I learned a lesson about our business: Never let a job start without having lien rights, which works in *most* cases.

The Aluminum Pipe Company

Another disaster in the '50s happened when I was a wireman on one of our service trucks. A fellow journeyman and I were assigned to perform maintenance work at a plant called Rollweld Pipe Company, where Miller had a long-term maintenance agreement. The company was in the process of expanding beyond their aluminum pipe production, to include rolling out big twelve-inch-diameter steel pipes using a completely new system.

Our job lasted several months and we logged twelve-hour shifts every day. It was tough to work seven twelve-hour shifts a week without a day off, yet we kept it up. After a couple of months our salary and overtime amounted to a lot of money. Ultimately the company lost too much money on the overall project and went bankrupt. The local financier who owned it said it wasn't profitable and he closed it, leaving Miller Electric with another unpaid bill. That was one more lesson on the obstacles of doing business: Although we had lien rights on the facility, we still couldn't collect at the end due to their bankruptcy.

Reorganization

Starting in 1952 and on to the early '60s Miller Electric passed through many changes, several very traumatic. The problems began with the death of Mr. Miller, who was killed in an automobile accident on December 7, 1952. I was an

1943 Citation for Miller Electric's War Manufacturing

apprentice at the time, so much of what happened would not be known to me until I became president. Suffice it to say, Mr. Miller's death set the stage for the future development of the company and many challenges.

The first order of business following his death was to reorganize. His wife, Nan Miller, was a nonworking member of the family who inherited Mr. Miller's stock. His daughter, Jane Wynn, was also a stockholder although with a minority share. The two women got together and decided to appoint Jim Dandelake as president. The action made a portion of stock available to him. Jim was the company's senior estimator who worked for Miller throughout the war, so he was the right choice for president. The result of those decisions was a company split into three parts: Jane Wynn, who received a third of the stocks; Mrs. Miller, who received a third; and Jim Dandelake, who would soon receive a third.

There was a fourth person in the mix, who would have a major impact on the company: Wesley Paxson, Nan Miller's son and Mr. Miller's stepson. Wesley worked for the company as an engineer and, while he didn't get any shares of the stock, he still figured as a key member of management. Apparently, Wesley felt he should have received a greater role in Miller, but since he didn't he decided to form his own company. In late 1955 he resigned from Miller and started Paxson Electric Company.

After Wesley made his move, his mother, Nan Miller, decided to leave the company to support her son in his new venture, so she sold her stock back to the company. But it didn't end there because when Wesley formed his new company he took several of our customers. He also hired away some of our engineers and estimators.

Obviously those moves created bad blood between him and his stepsister, Jane Wynn. You can imagine the impact it had on Miller … it split the company wide open. Friends had to part with friends to go work in competitive companies and many of our trade advantages walked out the door. Another problem was sensitive information belonging to Miller Electric had been compromised, making it extremely difficult to compete in the bidding process. We were affected in every way and it was a traumatic time, to say the least.

I didn't realize how bad it was until I moved into management and saw how things were. There was a real possibility the company would go downhill fast and,

Mrs. Wynn and her father, Henry Miller

as a matter fact, it did over the next several years. We dropped from a million-dollar Dunn & Bradstreet performer in the '50s, to an almost broke, high-risk organization in the early '60s.

Miller continued to struggle until Mr. Dandelake became ill in 1964 and had to leave the company. He officially retired in 1965, at which time Mrs. Wynn purchased his stock … a move that opened the door for me to be the next president.

By way of background, the Cecil Powell Bonding Company actually wrote the first bond for Miller in 1930 and posted it for Miller's work at the Federal Office Building in downtown Jacksonville. Their friendship grew and, among other things, Cecil helped Miller find work during the war. During our lean years, however, we had no bonding capacity at all due to our financial restrictions. So when Mrs. Wynn wanted to buy Jim Dandelake's stock, Cecil Powell gave her a check for $50,000 to help out and allow me to take over. It was more as a favor to a lifelong friend than a bond issue. Mrs. Wynn's husband also gave the company $50,000 for additional working capital. Those were non-interest-bearing loans, which show the kind of strong friendship Powell had with Dr. and Mrs. Wynn.

I was their only senior estimator with licenses to operate the company, so it was natural for her to ask me to be the president. Nevertheless, I was humbled by the high degree of trust exercised by Mrs. Wynn, Cecil, and others. That's why that morning in 1966 when I became president, I fully realized I had to keep their trust and move the company ahead however possible.

Courage and Passion: Tom Autrey

My father, Buck Autrey, *is* Miller Electric. He joined the company at the bottom of the totem pole as a helper, and over time became president, CEO, and majority shareholder. When he took over in the '60s, he had to bring the company out of near bankruptcy to become one of the top ten electrical contractors in the nation. He is a dedicated, intelligent, hardworking individual who represents the quintessential American success story.

I've been with Miller Electric for sixteen years and now oversee all maintenance, location tracking, and accounting for the company's fleet. With that kind of job I've been able to view my father from many different perspectives. There have been no surprises. He has always been straightforward with all of us in the family, and if you work for him—as most of his children have—you must carry your weight and get results. Period. And you can't make excuses when you don't get results.

I admire his work ethic, his courage, and the personality characteristics that made him a successful businessman. Among his strongest skills are his ability to successfully communicate, negotiate, and handle difficult people and situations.

My father also knows how to have fun. During the '70s on the rare occasion he wasn't working, he could be found at his lake house with a fishing rod in his hand. Mom and Dad later purchased a sport fishing boat, and Dad became a champion tournament fisherman. For many years they drove their RV to the University of Florida where I went to college, to attend the football games. Today my wife and I carry on this tradition. Once, in the early '70s my brother and I went in his office when he was out and I sat in his chair to see what it felt like. Suddenly my dad walked in on us and scared the crap out of me. I flinched and spilled my coffee all over his desk. I thought I was in deep trouble, but my dad and my brother only laughed it off.

I wouldn't change anything about Dad. He definitely became the person God intended him to be, which is all any of us can hope for. As for the best advice he gave me, I'd say it was to be unselfish. "This," he said, "is the key to happiness in life." I fully agree.

I'm glad to say my father raised me to be the man I am supposed to be. As a father, he conducted himself in a way that would nurture my spiritual growth and teach me to face life head-on, with my eyes wide open and my feet on the ground. He taught me how to act like a man when facing the challenges of life. He taught me there is a power within me that I must learn to tap into to achieve the greatness I was destined for. He taught me that you either go through life with courage and passion, or you retreat from life to wither away and die.

Over the course of my life I have made many wrong decisions, and I have probably not succeeded along the career path my father wished. However, I know in my heart that my father is at peace with himself when he sees me, for he knows that I have become the man I was destined to become. Through the good and bad events of my life, my father has been there the whole time watching over me. I am proud … very proud to be his son.

Tom Autrey
Fleet Manager

A Word about Delegation

In my experience, delegation of authority is the first thing we try to impart once a person achieves a certain level in their job. When a person becomes a project manager, for example, the worst thing anyone can do is micromanage him. Delegation is about trusting that employee with the job and giving them the freedom to do it. And along with that delegation they need the authority to accomplish the task, which gives them responsibility for the outcome.

Delegation in the team management context works when a manager and an assistant project manager, for example, work together for total management of the task. The team management concept with appropriate delegation is what we've always worked for throughout the company, and we still do.

THREE
The Survival Years

From the Bottom

As I mentioned earlier, I made a commitment to Mrs. Wynn to stay on as president even though the prospects for the future of the company looked very dim. The arrangement included stocks Mrs. Wynn would buy from Mr. Dandelake after he left the company, and the way it came about reflected the strong culture of trust and loyalty in the company.

Miller's bonding agent, Cecil Powell, had been a good friend of Jane Wynn and Mr. Miller for many years and Cecil was the only agent the company ever used. Without hesitation Cecil wrote a check for $50,000 and handed it to Mrs. Wynn to help pay for Dandelake's stock. The value of the stock at the time was fixed at $55,000, which doesn't sound like much today but was a lot of money at the time. So I accepted the option and gradually paid it off over the next ten years. There were some years when things were particularly tight and I had to borrow money from a friend to continue exercising the option.

***

Now that I was president, I could see the long-term effects of decisions and events in 1956 after the company broke up. For one, our spinoff company, Jacksonville Metal and Plastic that thrived since World War II, gradually went downhill and had to be phased out. The decline of our business, for reasons already discussed, pushed several of our prime employees to other companies. Engineers such as Phil Paille—a lifelong friend—left to work for the Paxson Company. Ken Stowe, one of my early mentors, along with several others left for Fischbach &

Moore, a main competitor. Then there was Ed Snyder, a good engineer I worked with at the Kings Bay project. Ed also left to form his own company, Snyder Electric. Many of our good engineers left us to go out on their own, leaving me and a few others to carry on. When I took over we had about forty employees, down from around 1,500 in the early '50s.

During those stressful times, it became known throughout the Jacksonville business community that Miller Electric had lost several good people. As a result, some of the healthy companies offered me a position, believing I'd jump ship like so many others. I have to admit I was tempted, but I couldn't turn around and leave Miller ... although it would have been easy to work for some other, larger company. I discussed it with my wife, but she was persistent and urged me to stay on. That was another case where I attribute most of my success to my wife. An interesting footnote is many of those companies that offered me a job have gone out of business.

When I took over, the company was heavily in debt. Among several other creditors, Miller owed the General Electric Corporation around $350,000, at a time when our net worth was about zero. The way that worked out is a powerful story in itself, which I'll talk about later. But at the time I found out the predicament we were in, I still had my doubts as to whether I made the right move.

Struggling

The year 1966 was quite a contrast to the early '50s when landing sizeable contracts was not unusual. The Savannah River project, for example, was a nuclear plant in Aiken where we had as many as 500 electricians on the job. The facility produced many of our original top engineers such as Ken Stowe and Ed Snyder. When the job was done, they and many others transferred to Jacksonville. At the end of the construction part of the job Miller secured the maintenance contract, along with the E. I. DuPont Company. We kept that contract all the way through 1990, when DuPont bowed out of the nuclear management business. But that didn't help much in 1966.

What helped keep us going in those years were several public school contracts I landed long before I became president. When the Baby Boomers flooded the

At a NECA event - 1962

They finally agreed it was a logical argument and decided to give me the letter of credit. It all had to happen since it was the only arrangement allowing us to go forward. In the end we took GE's credit offer and secured a line of credit at the Florida National Bank. To assure the bank we would deliver, we pledged our receivables each week in order to get the money to make payroll. It was a close-to-the-wire situation for a long time, and we even had a member of the General Electric Company attend almost all of our staff meetings. We definitely felt their presence, yet that was a good thing since they helped us put job proposals together … their way of assuring we made money to pay our debt to them.

Through the years of climbing our way back in the industry and paying off GE, Miller didn't accumulate much money at all. Every dollar we earned went toward paying off that GE debt. That's why I call the years between 1965 and 1972 "the survival years." We not only survived, we pulled out of the hole and started to rebuild the company.

It took us three years to pay off the entire GE debt. In the process I developed a close relationship with their top management that lasted many, many years. Long after our debt was clear, I heard about a story Reginald Jones told his staff at a luncheon meeting. He related how we had a lot of debt and that he took a chance and cut a deal with us one day at the Green Turtle Restaurant. "That young man," he said, "has paid us off in many millions of dollars' worth of business for the General Electric Company for years after that."

I count that lunch at the Green Turtle as one of the highlights of my career.

***

After our success paying off GE, I became one of Reginald Jones' fair-haired boys. As our relationship and business with GE grew, I was frequently invited to The Masters golf tournament as a guest of General Electric Supply Company. Each year a few of us would go up to The Masters tournament in Augusta, where I tried to play a couple of rounds of golf with Reginald. We'd play early in the morning and later on attend The Masters championship. Reginald was a novice golfer who sincerely wanted to learn the game, so I gave him a few pointers along the way. Helping him improve his golf didn't do any harm to our business relationship either.

school systems, Florida's Duval County put out a bond issue to build about thirty or forty schools. We aggressively pursued those contracts and, fortunately, we were successful in wiring the majority of them. Although we didn't make much money, the contracts paid our overhead through our most critical years. It also helped that the school projects, along with other new business, increased our sales volume and made our books look somewhat better. But we weren't out of the woods; Miller's outstanding debt, combined with competitive industry pressures, forced us to hone our budgeting and labor skills to a fine point. As things gradually improved, we re-hired some of the engineers we lost and also brought in some new blood, such as Ed Witt.

Ed came on board in 1957, and later, as an apprentice, attended my apprenticeship training during the evenings. I immediately recognized him as a bright student. Ed also worked for me as an apprentice while I was a journeyman for a short while, and we got to know each other well. Within a few years I had him inside working with me. Ed was not only a good friend; he was a very good electrician and foreman who developed into a good estimator. I remember the day we shook hands: "Ed," I said, "all I ask of you is to be loyal to me and support me in every endeavor we go into. If you do that, we'll be together for life."

***

Another problem we had in 1966 was bonding. We'd reached a point where we had no bonding capacity whatsoever, and that forced us into joint ventures with larger companies. Falling under the bond umbrella of larger companies became very expensive since we had to forego half of our profits to our partner company who signed for our bond. However, there was a bright side to that arrangement. I became familiar with several principals of those companies, such as the E. C. Ernst company, Fischbach & Moore, and the Dynalectric Corporation—one of the companies that asked me to join them. These relationships would pay off well in the future of Miller Electric.

Our final joint venture was with Dynalectric Corporation at Jacksonville's bulk mailing facility. It was a successful job and afterward we were able to secure our own bonding and credit capabilities. We also boosted our ratings that enabled us to launch our own financing, which led to a relationship with the Florida National Bank as well as solid lines of credit with our suppliers. By 1972 we were off and running, with no immediate problems as far as finances were concerned.

Eventually Miller began to land significant jobs. Contracts like the Barnett Bank building downtown, and the Blue Cross Blue Shield buildings ... all came in one right after the other. So there we were, six years after I took over Miller, and it looked like we would enter better times. But we still had a long, lean road to solvency.

A Word about Money

That's easy ... don't lose any! Or if you're losing it, cut your losses as quickly as possible. Managing a losing job is much harder than managing a *profitable* job. Some of the biggest mistakes estimators make (and I've made them myself) is when a job is going bad, to turn away as if to say, "There's not much I can do about it." The problem with that is every dollar saved on the losing job is worth twice the dollars made on a winning job. Put another way, you have to make twice as many dollars on a profitable job to offset every dollar lost on a losing job. That's a lesson many estimators can't get through their heads ... they have to spend a lot more time managing a losing job than they do a successful job. One of the best lessons I learned early on is to closely manage out the losing situations and not just turn my back on it.

You learn a lot when you lose money on the job. You learn how to cover your losses. One of the worst things an estimator can do is not admit they made a mistake and try to work around it or cover it up. That can be the most disastrous thing because someone else might have experience with the same situation from another job and they might help work it out. Remember: If you find yourself in a bad situation, take the problem to management and let them work out a scheme to cut your losses. While that's often a difficult lesson for an estimator to learn, it's the best way to handle it. There is no secret to making money in this business. All you have to do is eliminate the losses and that is done by creating good estimates to begin with.

Moving Ahead

We owed General Electric over a third of a million dollars, so if we were to survive as a company we had to get out from under that enormous debt. To remedy the situation I arranged to meet with the president of General Electric Supply, Reginald Jones, who would eventually become the chairman of the GE Corporation. So in late 1966 I met with Mr. Jones and a few of his colleagues at the Green Turtle Restaurant on Philips Highway. By the time lunch was over, I convinced them to not only set aside our debt on a non-interest-bearing note, but also to give us a million-dollar line of credit. I'm still amazed at how it happened … I guess I just bluffed my way through. It went something like this:

"Sure," I said, "you can write us off, close us down, and take a $350,000 loss. But …"—I pointed my finger at Reginald—"you just became president of General Electric Supply and that won't look too good on your résumé, will it?"

Seated next to Reginald was their new financial director, Ralph Glotzbock. He turned to Reginald and laughed. "You'd probably get fired over this, you know."

After they stopped laughing, I continued.

"Seriously, if you close us up, you'll probably do us a big favor because I'll turn right around and open up again the next day. But it won't be called Miller Electric … it will be under my own name and we'll continue in the electrical business."

No one said a word, so I pushed on.

"If you work with us, I'll promise you this: I'll work day and night until we get this debt paid off, and I will be a friend for life with General Electric Company."

In the end, Reginald Jones made a decision. He spoke to his colleagues. "I believe he is the right person for this job," he said, "and I believe we can work this out."

He turned to his credit manager, Red Gadbury. "You set up the note. Have him sign it, and we'll proceed from there. Oh, and give them the line of credit they need."

They discussed that last part about the line of credit, and Gadbury said he didn't want to give us one.

"How am I going to pay the debt off," I said, "if I don't have a line of credit? If I can't get supplies, I can't wire the buildings. And if I can't wire the buildings, I can't pay anybody back, now can I?"

One day in 1975 a few of us attended a convention in New York. At the time Reginald was busy in the GE headquarters in Connecticut, so he sent his helicopter down to pick up several of us at the Wall Street landing pad. Seated around his lunch table, Reginald told the story about how I gave him golf lessons. Then he pulled out one of his Connecticut golf course cards and proudly showed it to the group. The card had a 72 score ... obviously a good round and great kudos for me since I helped his game.

The way we got out of debt with the General Electric Company is a lesson on how people can work together to generate strong profits. I can also say the exercise helped us develop a stronger culture of efficiency that benefitted us for years to come.

Growing Again

We gradually increased our management team by bringing in new blood. Some of it came from our industry, but we mostly promoted those of our own who, like me, had worked in the field. We observed employees who'd been with the company for a while and watched how they conducted themselves with clients and other contractors. Satisfied they were good management material, we brought them in, taught them how to estimate, and exposed them to the Miller management culture. A few years later they became an integral part of the company. As always, developing people from within paid off with an expanding base of employees ... further proof it is less expensive to keep employees than hire new ones.

One of my early, best memories at the company happened when I dropped by Mrs. Wynn's office and told her we made our first million-dollar billing. That was in the early '80s and we finally reached the point where we could bill $1 million a month. That was a big deal at the time, and it eventually translated to $12 million per year. After we reached $2 million a month, it didn't seem like it was that big of a deal. Then we topped $3 million a month, and $4 million. It wasn't long until we didn't pay much attention to it. Instead, we paid more attention to how much *profit* we made. After we settled down from the turbulent years we averaged around 15 percent gross profit, and did so for many years after that.

Fifteen percent had always been our base number and we never wanted to do less than that, although we've done as much as 20 percent on occasion.

In 1972, Mrs. Wynn and I developed the company's profit-sharing pension program. At first, there were only a few people who participated and I believe we initially paid $40,000 into the fund. All the inside employees participated, and over time the fund developed into one of the most lucrative retirement plans that was ever developed. We paid into that plan over the years and continue to do so. Many people have shared greatly in it, and many have retired with a good amount of money ... retiring with dignity without having to rely on outside help.

A Word about Communication

Another strong point about the Miller culture is open communication. That's why every door in our Miller Electric offices remains open ... nobody works behind closed doors unless they need privacy to work on a bid or something. Everyone can walk right in without having to knock if they want to speak with a peer or someone else. I don't make appointments and if anyone wants to talk with me they know my door is open. That's always been the case. All employees, from warehouse clerks to accounting clerks, estimators, engineers, and so on, know they can sit down with me to discuss any problem and know it won't find its way back to their superiors.

Of course we've had some communication problems in the past. Most of those were brought about by competition among different groups in the company. Not that there's anything wrong with competition. In fact, I promote competition because I want our people to be better. "Look how he's doing much better than me" ... that kind of competition is a healthy way to find out why, and make themselves better. It's perfectly all right to expose who's the best or who's the next best, to let them know where they stand. With that kind of awareness and openness, they are encouraged to communicate even more. Sometimes closed communication can develop a mentality of turf building. "I don't want to talk with so-and-so because ..." or, "If they find out I'm doing such and such, they may want to move into my territory ...," and so on. We've had our share of that kind of closed-mindedness, but we're beyond that now. In short, open communication and transparency is the best way to move ahead.

And another thing: When I was president we had a one-on-one system where I made it my goal to visit each person in the company from time to time. I did it to get a personal sense of how they felt about the company, how they got along with their peers, and if there were problems I needed to know about. It worked well, and today we have so many employees we have to get most of our feedback through an electronic suggestion box. We recently completed a confidential online poll where employees posted questions and other employees offered solutions. We had a lot of excellent comments, so we'll probably do it again. So, successful relationships—whether at work, home, or anywhere—begin with good communication.

FOUR
The Autreys

Where Credit Is Due

During our first year of marriage Betty took a job at a commercial store in downtown Jacksonville. That didn't work out, so she found a job at a local kindergarten named Kiddie College. Her role was to drive some of the kids to and from school in her English Ford—the same one we used on our first date. Between driving duties, she served as a teacher's assistant in that kindergarten.

Our own kids came on the scene fast. Like doorsteps ... *one-two-three-four* ... they arrived one right after the other. Our first daughter, Susan, was born in February 1952. In December of the same year our oldest son, Ronnie, was born. His is an interesting story I'll tell you about shortly. Our second daughter, Sandra, came into the world in 1954, followed by Tommy, our youngest, in 1956.

After our first child arrived, Betty continued to teach, and when the second one arrived she sort of tapered off. Although she no longer worked at Kiddie College, all our kids attended that good institution through their preschool years. As it turned out, that position would be her last outside job. She decided that with four children she would dedicate her life as a full-time mom and homemaker. I'm glad she made that choice, and she's done a wonderful job.

During her early childbearing years, I attended school four nights a week and worked all day, every day. I often had to leave at five o'clock in the morning and wouldn't get home until eleven at night. A lot of times I'd leave work and go directly to school, changing clothes on the run. So I didn't have much participation in the early years with our kids, until later on. Fortunately, my wife was always there. That's why I can say she has been my greatest partner, my greatest friend, and my greatest supporter in everything I've done. As time passed and our kids grew into adults, I can say the same thing about them as well.

***

The hours we had to spend at our jobs made balancing work and family difficult at times. When Ed Witt and I worked together in the early stages of Miller, it was normal to spend twelve to fourteen hours on a job every day of the week. We tried to be at home with our families on weekends, although we didn't always succeed. *Most* weekends, however, were dedicated to our families.

When our kids were in grade school, we moved out to the beach where they attended Sea Breeze Elementary. Around that time, music became a significant part of our lives. All four of them found themselves in some kind of musical activity, so Betty and I encouraged them to join the school band. They did, and we became "band parents," which of course led to joining the Band Parents Association. In addition to their musical activities, Susan and Sandy also made the Majorettes squad and Tommy became a drum major in his senior year. I guess it was in their genes since I always enjoyed music. I used to play the guitar and sang occasionally. We attended all their concerts and did the things parents do with their kids during high school.

That included many camping trips. I bought a small pop-up camping trailer when the kids were in their teens and we spent many weekends in various parks around Northeast Florida. It was a good way to have quality family time, and Betty enjoyed it particularly when it came to food. She was really good at preparing delicious meals on the gas stove in our trailer. We kept it up for a few years, until I sold it right before Hurricane Dora visited us in September of 1964.

Dora was the last real hurricane to hit Greater Jacksonville and she knocked many sections of the city out of power for over a week. Our house was all electric so that had a major effect on a lot of things, particularly my wife. She jumped on my case that week because when I sold the pop-up trailer I also got rid of her three-burner gas stove. For several days following Dora, Betty had to cook over the fireplace … and boy, did she cook! My brother stayed with us at the time and he raved about her impressive meals in spite of having to cook like a frontier wife.

Dad and Grandma, Lake House 80's

The Children

It is a wonderful thing when your whole family is involved in the business, and that's exactly how it worked out with my wife and children. I already told how Betty worked alongside me through my training days and onward, so when the kids came along we raised them to know they had to earn their way. And they did. At one time or other all four of them worked at Miller Electric.

After Susan graduated from high school, she found a job at an insurance company and stayed there for a couple of years. Later in 1974 at the age of 22, she went to work for Miller Electric and has been with us ever since. She was a dedicated employee in the accounting department, working directly under Mrs. Wynn who mentored her on everything about accounting. Eventually Susan took over that department, and in 2004 become CFO of Miller. She handled the finances for the company, including key financial arrangements with the bank. Now, of course, she's part owner of the company, along with her two sons and our employees through our Employee Stock Ownership Plan.

***

December 28, 1952, marked the beginning of a critical time for Betty and me. Ronnie was born on that day, and within three days of his birth it was obvious he might not make it. The problem was he couldn't get enough oxygen into his system. The doctors called him a "blue baby," a *cyanotic* child whose condition fails to oxygenate the blood. The doctor kept him in an oxygen tent for several days, but it didn't get better.

Then Dr. Skinner, our pediatrician, called me to his office. He explained he should perform a rare surgical procedure to withdraw the air from around Ronnie's lungs because they collapsed moments after he was born. The doctor told me a speedy operation was critical. It would be a new procedure where he'd use hypodermic needles to extract the air and make the lungs expand. He said he needed my permission to proceed. I agreed, of course.

My wife was not aware of what happened, but only knew she hadn't held her baby and was very upset about it. She also knew Ronnie was premature, which

RV to the ball game

The four doorsteps Ronnie, Susan, Sandra and Tommy

In the early 1960s

Date night

was the reason they kept him in the oxygen tent. The more time passed, the more upset she became and each day had to be an eternity for her. The doctor and I thought it best not to tell her about the situation, so I prayed a lot.

After the surgery when we knew everything was okay, a nurse brought the baby to my wife. It had been a week or so before she actually held her baby! That was when I finally told her what happened, and how much I hated keeping her in the dark. Three weeks later the doctors decided Ronnie could breathe well enough on his own, so we took him home. Ronnie had no more problems after that. He recovered and grew up to be a big, strapping boy.

Later in school, Ron went into sports and became an excellent pole-vaulter who earned an athletic scholarship at Western Carolina University. He only stayed a couple of years because "It was too hard to pole-vault in the snow." He called us and said he wanted to come home, which he did. Back in Florida he enrolled in Stetson University and later transferred to the University of Central Florida to pursue his engineering degree. But he apparently ran out of steam there and decided to work for Miller. He eventually earned his degree at Jacksonville University.

Ron started at Miller as a summer field and warehouse helper. After several years and working with nearly every estimator in the plant, he became Miller's president in 2004. When he took over we grossed around $90 to $100 million in sales. The boom years moved us along with a steady upward trend, starting at $100 to $125 million, then $150, $200, $250, and $300 million. It was a period of explosive growth when our biggest year topped over $330 million.

In addition to the increased sales volume, Ron generated far greater profitability than ever and doubled the net worth of the company. As a result we paid good bonuses to our employees. It was an impressive run and I have to say that Ron did an excellent job as president.

A True Patriarch: Ron Autrey

My first job with Miller Electric Company was in 1967 while attending high school. I worked in the warehouse, sweeping floors and taking inventory. Later in my senior year and throughout the summer of 1970 I worked in the office counting electrical devices and fixtures on drawings. In 1973 I enlisted in the Army for a four-year commitment. Following basic and advanced training, I was assigned to the Army Security Agency, specializing in electronics maintenance.

As the war in Vietnam wound down and thousands of troops came home each week, I was given the opportunity to leave the Army early with an Honorable Discharge. Their proviso was I had to prove potential employment when I got out. With undesirable pending orders for Turkey or Alaska, I called Dad and explained the situation. With the Army's criteria satisfied, I left to join Miller Electric in November of 1974.

After I began full-time employment, my dad asked me what I thought I should be paid. When I began to tabulate the costs of my new life, he interrupted me. "You do not get paid what you need," he said. "You are paid what you are worth." Sometime after we established my pay, I learned another quote from him: "Always pay them a little more than they think they are worth."

I began my career working as an office assistant counting electrical devices and lighting fixtures for pricing estimates. As a junior estimator I worked for several different project managers and vice presidents, and gradually assumed more duties and responsibilities. In 1979 I became a project manager and later a senior project manager. As a primary qualifying agent, I was licensed as an electrical contractor in twenty Southern and Midwestern states and also as a general contractor in the state of Florida.

When I was the service department vice president in the mid-'90s, I led revenue growth in a four-year period from $2 million to more than $23 million, and become senior vice president. As president and later

CEO, our revenue grew from $96 million in 2002 to more than $311 million four years later. Gross profits had risen from $17 million to more than $50 million in the same time period. In 2012 I was named chairman of Miller Electric, and Dad became chairman emeritus.

None of my business success would have been possible without the opportunities provided to me by my father and Miller Electric Company. He did not meet with me often and he rarely held group meetings. He was not a micromanager and was always available to give advice. I learned quickly that if I took a problem to him, he was pragmatic, a quick study, and swift to lay out solutions and action plans.

Dad's impact on Miller Electric is ubiquitous. He is widely known and respected for his leadership both here and at NECA. Among his many accomplishments was preventing the abolition of expanding the National Electrical Benefit Fund (NEBF) in 1980. As NECA president, Dad presided over a controversial vote of the board of governors to increase the benefit contribution to 3 percent of gross wages. As a result, today's NEBF pension assets total more than $12 billion, providing retirement benefits for more than 536,000 electrical workers.

Dad's dogmatic practice of reinvesting profits into the company supported employee rewards, company development, the implementation of technological innovations, and internal financing for continuous company growth. He also created cash bonuses and profit-sharing retirement programs, along with open-ended opportunities for advancement of employees. With these innovations, employees could prosper, knowing their careers could start and end at Miller Electric.

Dad is a true patriarch, with a span of management and leadership talents extending beyond the basic control of a company and family. He is a calm, decisive leader in times of crisis. He is slow to anger and never raises his voice. He has no propensity to use foul or offensive language. He does, however, speak the truth, at times even if the truth is painful. While he has no advanced listing of academic degrees, many accomplished people say he is the smartest man they know. I am also one of those believers.

Dad is serious about his profession and passionate about his family, but he can also be funny. Once he delivered what I call his best line ever. While hosting a dinner party at his home for the office employees and senior field managers, he offered to buy his friend and field superintendent, Marvin, a drink at the family room bar. Marvin explained that he did not drink, and my dad said (I'll never forget it): "Well, you *should* drink so people will think you're drunk instead of just stupid."

Dad is generous and never greedy. He has a heartfelt appreciation for his own success and has always been willing and motivated to share with others. He does so not for self-recognition, but to enable others to get a step up in life as they pursue their own goals.

Dad is a good listener. Even when his plate was full and his own tempests boiled, he had the ability to stop what he was doing and listen. However, he was not patient. After he heard your story, you had better be ready for action.

Dad was a good teacher. He had innate skills as a craftsman and he didn't mind sharing his knowledge through demonstration and assistance. As a business leader, he guided many people toward successful outcomes with his sage advice.

Dad is a good father. He was a backstop for his children's development and our failings. He is never outwardly judgmental and always willing to step in and stand up for his family.

Not enough words can be written here to document the sometimes gruff but loving support he has provided his family. When we were children there was never a fearful moment caused by his anger. However, we did fear that we might not measure up to the exemplary standards and work ethic that he demonstrated throughout his life.

One day in 1980 Dad said, "Don't worry about making money, just don't lose any!" What he meant was he had built a company with a consistent operating model that produced consistent results, and as long as we didn't change the model and continued to follow best practices, the profits would be there.

To say that "Behind every great man, there is a great woman" is understating the love, support, and motivation provided by my mother, Betty Maxine White Autrey.

Dad and the children, grandchildren, and great-grandchildren are a living legacy to the love, pain, and sacrifice my mother provided without hesitation or discrimination. She was there for the good and bad times, with love and support.

She held the line at what was moral, right, and just. She demanded our genuine honesty and respect for others, and the conventions of what it meant to be a good person. She waved the flag at every finish line, and praised the accomplishments of every one of us. She did not judge, and she did not waver in her relentless support of the work and challenges we faced.

My dad could have never accomplished the things he has done without the constant love and support that she has consistently provided my father for the last sixty-five years.

Ron Autrey

Chairman, Miller Electric

Our daughter, Sandy, and her husband, Buzz, were high school sweethearts who married right after graduation. Buzz already worked for Miller and eventually became a project manager for the paper industry and, later on, a vice president. During those years Sandy always wanted to stay home with the kids, so she worked part-time at the company. Later she moved with her husband to various job locations in Pensacola, Florida, Plymouth, North Carolina, and Augusta, Georgia. They eventually returned to Jacksonville until Buzz retired.

Eventually, Sandy came back to Miller to work as a CAD operator, and she is now manager of our BIM (Building Information Modeling) department. She and her team play a key role in our pre-construction process.

***

Tommy was a model airplane enthusiast who spent weeks on end building model airplanes. After countless hours laboring over a new plane, he'd take it out to the strip with his mother, start it up, and let it fly … only to have it crash in no time. That was okay, since he went right back and started over again.

Later in life, his love of model airplanes led him to become an excellent pilot and flight instructor. He was also an accomplished guitarist and singer, which I'll discuss in a moment. Tommy joined Miller in the '90s and grew into an excellent technician and coordinator. He continues to work with us to this day, managing our fleet department.

In Quiet Times

My wife and I are both prolific readers, poring through a half-dozen novels or so each month. There are lots of authors we follow, such as Lee Child who wrote the Jack Reacher series, and the Doc Ford Adventures about Southwest Florida by Randy Wayne White. I also read the Bible a lot, mostly the Old Testament stories. I'm more interested in the stories than I am for its religious aspect, though, since I'm not a real religious person. On the other hand, I'm not an atheist either. I simply don't get into religion too much, that's all.

Aside from the Bible and certain novels, my favorite books are adventure types and mysteries by authors such as Earl Stanley Gardner. He's one of the best mystery writers out there, in my opinion. And there's John D. MacDonald, a truly prolific mystery writer with around eighty or ninety books in print, and Wilbur Smith's books on the cradle of civilization in Africa. I think I've read every one of them. Lately I've been reading current books that come out on the bestseller list.

As for movies, well, I haven't been inside a movie theatre in probably thirty years … mostly due to my bad hearing that makes it difficult to understand. Betty and I watch a lot of sports programs like football and golf, and old country-western movies. When you talk country-western, I include the action westerns, or others like the one with that great line, "Go ahead, make my day." I think I've seen every one of Clint Eastwood's movies at least twice. Eastwood is one of my favorite stars, along with John Wayne, of course, who had to be the most prolific actor in history. Wayne didn't quit until he had one foot in the grave, and

Eric's wedding, Atlantis, Bahamas, April 2010

Eastwood will probably wind up the same. He's still directing some of the best movies out there today.

My favorite music comes from stringed instruments. In the evenings at our home in Palatka, Florida, Betty and I will have cocktails on the deck and listen to DIRECTV's *Wonderful Instruments* program. They play popular string music on the violin, cello, guitar, harp, and the piano, which is also a string instrument. As for radio music, give me country-western—probably my favorite. After all, my wife and I first danced on roller skates to the strains of country-western music.

The Crooner

In the early '60s, my wife bought a guitar for me with ten books of Green Stamps she diligently saved over many months. One day after work, I picked it up and started playing. I didn't really know how to play, so I just taught myself by ear. I often played it when camping with the kids, sitting around the fire. I'd pluck out a few songs like "On Top of Old Smokey" or "My Bonnie Lies over the Ocean." I'd strum the chords, mostly in the key of C, which was all I knew at the time, and the first thing we knew, most everybody in the camp came around the fire. They chimed in and harmonized like a regular community sing-along.

Once when I was vice president of NECA, I attended the annual joint Executive Committee meetings for NECA and IBEW (International Brotherhood of Electrical Workers). It was a business meeting where we approved changes in the use of funds, discussed labor issues, and so on. That particular year we met in Bermuda at the Hamilton Princess Hotel. Following a particularly long day, we gathered at the nightclub for dinner and cocktails where a small string band entertained us. After the band took a break I hopped onto the stage, picked up one of the electric guitars, and started strumming. The first thing you know we were singing along like we did around the campfire. Then people got up and started dancing.

When the band returned from their break, Charlie Pillard, the president of IBEW, tucked fifty or a hundred bucks in the bandleader's palm and told them to take another break since we were having much more fun. For the next hour I played guitar while they sang and danced ... just one of those things you could never do on purpose.

I continued to strum the guitar at home, and eventually my sons Tommy and Ronnie picked it up. They both became proficient and we eventually formed a little combo, even playing for Sandy's wedding reception at our home at the Isle of Palms. I tell you we were *the* entertainment that night. Today, Tommy is one of the best guitar players I've ever heard. Ronnie, too, did well with it and he still plays the guitar every now and then.

Over the years my guitar became a real escape. During Miller's developing years I would often come in late at night when the kids were already in bed. I'd quietly close the door, grab my guitar, and play—just me and the dogs. I'd pick away and sing songs as quietly as I could until one of the mutts started to mimic me. When I hit a high note, the dog would rear up on his haunches and croon right along with the music. It was fun and a good way to wind down after a long, tough day.

Overcoming Stress

There were many things to worry about in the electrical business, particularly in the earlier years of my presidency. I have to say I did lose a lot of sleep in those years, mostly from things I couldn't do anything about ... things beyond my control that caused more problems than anything else. If there are problems I can solve, I find a solution—good or bad—and move on. I don't count those as stress situations, just things that should be taken care of in the normal course of business.

What's more important is keeping the stress of work separate from my family. Throughout my career, I've been able to do that ... to set matters of work apart from family time. If I wake up at night with something on my mind, I get right up and make notes on how to solve the problem the next day. That way I can forget about it and go back to sleep.

I think I've taught myself to compartmentalize fairly well over the years ... to keep problems away from what I'm involved in at the moment. Then there are diversions that help keep me relaxed. I found playing guitar, golfing, fishing, and spending time with family all helped me keep balanced. Stress has to be the biggest killer out there, based on what I've seen in many businessmen. I've noticed many of them can't get rid of the stress or the bitterness of failure and don't know how to let it go.

Another thing that helps is delegation. If I had a problem, I delegated it to someone else who was able to deal with it and let them solve it. Of course, if they were unable to solve it I was right there to help out. And if I couldn't delegate it for some reason, I'd solve it myself by working through the problem until I came up with a solution, and moved on.

Sometimes decisions are tough, particularly if it's a loss situation on a particular job. When that happened, I cut my losses as much as possible and tried not to dwell on things I couldn't solve. Sure, there will always be losers and winners, but the idea of course is to have more winners than losers … preferably a lot more.

I never looked to the Lord to help solve my business problems because He has enough problems of His own. Nor do I belong to any particular religion, although I was baptized at the First Baptist Church in Jacksonville Beach. My wife and I have attended services at that church over the years and also St. Mark's Episcopal or a local Presbyterian church. Although we're not really churchgoing people, I've always believed the closest I can get to the Lord is on the water. Heading out to sea early in the morning under a beautiful sunrise and seeing awesome occurrences of nature makes me believe there is a greater Being somewhere who makes this happen.

Of course, some things are so big and awful that stress cannot be ignored, such as what happened on 9/11. Anyone old enough to understand what happened on that day will remember exactly where they were. I do. A few minutes after I sat down at my desk that morning, my wife called and told me two airliners crashed into the World Trade Center's twin towers. I immediately turned on the TV and watched that terrible, terrible tragedy. It was a moment of national crisis that affected everybody alive enough to think.

What was interesting is on the business side of things; it didn't have much of an effect. The markets immediately dropped and although the travel and tour industry took a nosedive, electrical contracting proceeded as usual. We were involved in a lot of projects at the time and it never seemed to slow down. Eventually the economy rebounded and housing development fired up again, but commercial building never missed a beat.

Finally, good health is a big stress reliever. Fortunately, my genes go a long way to keep any potential health issues to a minimum. Otherwise, I try to stay in shape. Betty and I have always exercised by walking, jogging, and not eating too much. Keeping as fit as possible definitely lowers stress.

A Word about a Positive Attitude

Anyone lacking a positive attitude should get out of the business. We've lost so much money and so many jobs over the years that most other companies would've gone broke. But we didn't. At the time I took over the company we were down to forty employees and the only direction we could go was up. To head in that direction we needed to change our attitudes and forget what happened in the past. We did and we kept moving forward. A positive attitude tries to find ways of doing things better. When mistakes are made along the way, the positive attitude says mistakes can be overcome. We all made mistakes—a lot of them. But we survived, we grew, and we prospered. You will make mistakes too: mistakes in bidding, mistakes in the way you manage a job, mistakes in the contractors you select. You will make lots of mistakes, but how you overcome them makes the difference between success and failure. You can't let one job bring everything down … you have to offset that job with a better one. The positive attitude knows that, and seeks the best solution.

Giving Back

Research says the act of giving is a way to decrease stress and improve physical health. I agree, along with the fact that giving is simply the right thing to do. Throughout my career my wife and I have always contributed to charities in one way or another. Our involvement has increased as the business grew, whether through our own individual donations or with some Miller Electric involvement.

One charitable organization was our local USO. I was originally selected to serve on their board of directors, but through a series of events wound up as their president … which always seems to happen. Early on, I formed a committee composed of two people—an officer from the Prudential Insurance Company and me—to raise funds for a new USO facility at the Navy base in Mayport. At the time there was no USO at the beaches for sailors on shore leave, although there were tens of thousands of Navy men and women in the vicinity. We spearheaded the effort and within a year raised about $1 million.

When the new building was done I got a few Miller employees to volunteer their weekend time to do the electrical work. Bill Gothard, a Miller vice president, headed up the project. We also provided the materials and necessary financing for the entire project. It turned out to be a good accomplishment, and we received a lot of praise from the USO.

During that time I got to know some of the Navy personnel in the area. Kevin Delaney, a retired admiral, now deceased, was the commander of the US Naval Air Station in Jacksonville. Also a member of the USO, John involved me in many activities with the Navy. He also helped me get a Presidential Award from President George H. W. Bush, honoring me for my assistance to the USO.

The Boy Scouts of America is another favorite charity. One year I was awarded the status of Honorary Eagle Scout by the downtown Jacksonville Rotary Club, as a result of my contributions. Later when they invited me to serve on their board, I had to decline since I was still president of the USO, among other things. I didn't have the time to be elected president of yet another organization, so instead I included them on my permanent charity list.

Every year Betty and I sponsor the Vision Is Priceless BBQ, a local Jacksonville organization providing free screening to children for glaucoma and eyeglass needs. We also give to nearly all of the hospitals involved in charitable work, such as St. Vincent's. Last year alone, St. Vincent's donated over $85 million to charity care and nonprofit partners. Another is Baptist Hospital, which we've always supported. Of course, the Mayo Clinic Hospital is a major one for us and we support their foundation regularly.

We also sponsor one of the largest annual bass fishing tournaments in the nation. Based in Palatka, Florida, between 400 and 500 participating boats head out on the water to catch the biggest bass. All of the proceeds go to Baptist Children's Hospital in Jacksonville.

In the early '90s we made substantial contributions to Rotary International for their drive to eradicate polio throughout the world. Rotary solicited money

USO dedication at Mayport

Certificate of Award

The Secretary of the Navy

takes pleasure in presenting the

DISTINGUISHED CIVILIAN SERVICE AWARD

to

Henry E. "Buck" Autrey

in recognition and appreciation of the
distinguished services set forth in the following

Citation

For outstanding service to the Department of the Navy from January 1986 through December 1996, while unselfishly serving as a volunteer on the board of directors of the United Service Organization (USO) of Jacksonville, Florida. In 1988 and 1989, Mr. Autrey served as Chairman of the construction committee for the new $1 million USO center built near the Naval Station, Mayport. Through his personal efforts, over $200,000 of in-kind supplies and construction services were provided at no cost to the USO, ensuring the center could be completed free of debt. As the USO Council President in 1990 and 1991, Mr. Autrey guided the council into good business and accounting practices. He also established an endowment fund totaling over $250,000 to ensure continued service to military families. Through his continued service on the USO Council and in other community forums, he ensures that Navy and Marine Corps families in Northeast Florida and Southeast Georgia feel welcome and are properly served. Mr. Autrey's distinctive accomplishments and dedicated service reflect great credit upon himself and the Department of the Navy.

John H. Dalton
Secretary of the Navy
18 June 1997
Date

Secretary of the Navy Certificate

from their million-plus worldwide membership and raised close to a billion dollars. As of today we have been effective in eliminating polio from everywhere in the world except Pakistan and Afghanistan. Some areas were difficult to get into, such as Africa, China, and India; however, today those areas are free of the debilitating effects of polio. Most of the success is due to funds donated by the Rotary Foundation.

In all, we've made more contributions than I can number. While that seems like a lot, it isn't, really … not when I consider how fortunate we have been in our lives. I only point out some of the charities we support because I believe everyone should be involved in some way. It's the right thing to do, and good for the giver as well.

Company and Family: Susan Walden

I was in my early twenties when I came to work for Dad, and that was about forty-one years ago. After working with him all that time I still think my father is one of the most amazing people I have ever met. He is generous, particularly with his children and grandchildren, and he's one of the smartest men I know. Working with him has been the most wonderful thing in my life. He's so smart … I think I learned more from him about our accounting procedures than I did from the accountants we hired! He's an amazing person and I feel I'm so lucky—especially now—to still have him right next door to my office. I'm also fortunate to have my two boys working with me.

When it comes to working with my father there isn't much I'm surprised about, except we're still here after forty years and he still enjoys coming in every day. He might not be as involved in the business now, but Miller is as much a part of his life as it is for the rest of us. Still, I never would have thought forty years ago we'd be doing this today. And if I'm surprised to have been here this long, I'm not the only one.

Dad has been here over sixty years, and his partner, Ed Witt Sr., has been here almost as long. We also have many other people who have been here thirty or forty years. During our appreciation luncheon last year we gave over thirty awards to people with more than twenty-five years in the company, proving this is a great place to be.

I came in the early '70s when the company was already over the hump of the transition problems they had in the late '60s. It was through Dad's and Jane Wynn's diligent work that this company succeeded. If it weren't for Dad, I don't think this company would exist.

I don't know if I can point out a *single* favorite bit of advice from Dad … there were so many of them! I think being here, observing him, and letting his work ethic rub off on me (which I passed on to my sons) is one of the more important ones. To me, that is his legacy … his dedication— to the company and to his family—is beyond compare.

As for where he would be if not at work? Certainly in the last few years he'd be on his boat, but it wasn't always that way. Dad worked constantly during the day and taught apprenticeship school at night. Aside from that, he loved to go fishing, so we'd often go to the lake house. Mom and Dad loved going down there and they actually lived there for a good while, driving between the office and Lake Geneva in Keystone Heights. They loved to spend time there and now have their river house in Palatka.

When we were small, it was different since he worked a lot. I remember my mother telling me once, that when I was six years old Dad came home late one night and woke me up. I said, "Mama, who's that man coming in the door?"

As we grew older and had families, my parents kept expanding their lake house so each one of us had our own family suite for our spouse and kids. There was always a crowd of people at their house.

I can't think of anything I would change about Dad, unless I could make him younger so he could be here longer. As for my mother, she's always been his biggest supporter, and he's always been her biggest supporter. They have a strong relationship, and recently celebrated their sixty-fifth anniversary with a party at my house in June. I think they are a tremendous inspiration and role models for every one of us. Overall, Dad has been responsible for my success and the success of everyone in my family. And for this I am eternally grateful.

Susan Walden

Executive Vice President, Secretary-Treasurer

Work Like You're Never Going to Die ... Play Like You're Gonna Die Tomorrow

Throughout my life I've always had access to the water and for most of our married life we've lived on the water someplace or another. Whether on the St. Johns River, the lake in Keystone Heights, or at the Isle of Palms near Jacksonville Beach, nearly every home we've owned has been right on the water. Of course along with all that water came fishing, which I've enjoyed my entire life.

My wife's father was a fisherman, and I'll never forget the time he and his friend took me offshore fishing the week before I married Betty. We left St. Marks, Florida, and headed into the Gulf of Mexico where we fished most of the day. It was the first time I'd ever been out in the ocean and I got really sunburned. By the time our wedding day approached I'd reached the blister-and-peel stage, which greatly upset Betty. She peeled the skin off my ears and forehead and grumbled about how terrible I'd look for the wedding. "I didn't marry some ol' fisherman, did I?" she said.

When the big day arrived she was in a far better mood. That was my first experience with saltwater fishing, and it's been a love affair ever since.

Jimmy Haddock (his real name) was a talented fisherman as well as my journeyman during my apprentice years. We didn't have big boats then, so a lot of times we motored straight out from the St. Johns to the open sea in a rowboat, with nothing more than a small outboard and a set of oars for backup. We often caught all kinds of fish, like flounder, kingfish, sea trout, and sometimes shark.

Later in life I was inspired to get into sport fishing through my grandkids, Henry and Daniel Brown. It started in 1987 when I chartered a small boat out of Mayport to take them deep sea fishing. We powered to the mouth of the St. Johns River, and about an hour later came across a school of right whales. You can imagine it was one of the greatest things for young kids to see. There were several right whales in front of the boat, meandering around all slick and beautiful as they blew out breaths of condensed air which look like immense spouts of water.

Later on we switched to bottom fishing mode and caught several groupers, bonitas, and some sea bass. When it was about time to head back we saw the greatest sight of all: a huge orca whale. It mushroomed through the surface, shot up and out of the water, spun around, and fell on its back with a tremendous splash. It happened again and again, and we were only about ten miles offshore. I remember what my grandson, Daniel, said to my wife after we tied up at the dock.

"Nana, this is the greatest fishing day of my life. I'll never forget it!"

That's all it took. Right after that we bought a 44-foot Ocean Yachts fishing boat. It was a used boat, but the Ocean brand was well known and respected, and a perfect way to try out our new love for deep sea fishing. Daniel also took up fishing and he became my biggest inspiration to continue the sport myself. He came with me almost every time I went out. At first he was shy and stayed on the bridge with Betty and me, until the day she got him motivated.

"Come on, Daniel, let's go down there." Then she turned to the mates. "All right, guys, the next fish is Daniel's."

He was twelve years old when he caught his first big fish—a kingfish—and his catch placed us fourteenth in the 1993 Jacksonville Kingfish Tournament. I also recall the time Daniel caught his first blue marlin. He was around thirteen and we were right off St. Augustine. It was a thrilling day indeed, and after that his confidence in fishing soared. Whenever he was on the bridge with me and heard the *click* of the gear, he'd bolt down the ladder quicker than anyone else could get out of their seat.

***

A few seasons came and went, and then it was time to get a bigger, faster boat. We sold our 44' Ocean in 1991 and bought a 55' Ocean. We fished that one for about a year and a half and replaced it with a 58' Ocean … by far the best Ocean

At the Conch House after a day of fishing

Eric's First Marlin

Buck's Bahamas Billfish Championship 2004

Another successful tournament

Yachts Inc. ever built. Boat-wise, she became the love of my life. Over the next six years our fishing team won many tournaments on that boat. We caught quite a few blue marlin off St. Augustine, and won Boat of the Year in 1995 and 1996 from the Northeast Florida Marlin Association. We also earned runner-up twice. Eventually we expanded our horizons to the Bahamas and won a few tournaments there, primarily the Treasure Key Billfish Championship in the Abacos. Daniel fished those tournaments with me almost every year, helping us earn runner-up boat twice. In 1999 we won the Treasure Key tournament on points, and in 2000 Daniel caught a 620-pounder, to win the same tournament again. By the way, our 620-pounder was the first blue marlin we ever killed, but we had to because it was a money tournament and those were the rules. I'm glad to say in all the years I've fished, we've killed only six blue marlin.

It has been said that as soon as a boater takes his first cruise he'll start planning his next, bigger boat. It's true, and that's what most boaters do. They call it *two-foot-itis,* where two more feet of boat length will satisfy the longing to go bigger. My main reason to buy up to a larger boat was also versatility: a larger boat allowed us to take longer trips. That's why, after we won a few tournaments in the islands, it was time to get another boat.

I sold the 58' Ocean and bought a 60' Ocean, which turned out to win more tournament money than all the other boats we owned. But eventually, two-foot-itis set in again and I bought a 62' Ocean that I kept for six or seven years. We sold it in 2008 and switched brands to a 60' Viking I also named *Bet-A-Buc* (nearly all my boats had that name), and ran her for the next three years. Reducing the size indicated a certain design preference and also marked the beginning of the end of a long, enjoyable sport fishing run.

By then I'd been in the sport fishing circuit over twenty years, so my wife and I decided to give it up. Around that time Daniel had quit fishing with me after he married and moved to Mississippi, where his wife pursued her doctorate in psychology. When she graduated they moved back to Jacksonville and Daniel went to work with the company. But there was another reason we quit. After Daniel left, I had to hire a crew to fish with me, but it wasn't the same. I knew that was that. There was to be no more fishing for the *Bet-A-Buc.*

We sold the Viking in 2012 and bought a used Marlow cruising yacht, which we fell in love with. Since then we've ordered and taken delivery of a new Marlow, which we plan to cruise up and down the Intracoastal Waterway. As of today I estimate I've spent upward of $9 million on our boats, notwithstanding the trade-in values. Yes, boating is in our blood—a far cry from our early days on roller skates.

In Retrospect

For a fisherman, the greatest thrill in the world is to see a blue marlin hit your line, blast through the surface, and dance on its tail across the water. The adrenaline is such a rush you almost lose your breath just watching it. Another rush came when I'd back the boat down in a five-foot sea to help land the fish. Then the entire cockpit was awash with saltwater as the angler and crewmen got soaking wet trying to catch the big fish. It was a tremendous thrill and we enjoyed every moment.

Much of my joy in the sport was my view from the bridge, not in the cockpit. I was the captain and my job was to run the boat with the belief that if the angler didn't screw up, the boat would catch the fish. It's the *maneuvering* of the boat that makes or breaks the landing of a billfish … which is the reason I didn't catch many fish myself in my years of sport fishing. Probably the biggest one I ever caught was a sixty-pound kingfish.

Betty was truly my best fishing buddy … the admiral of the boat who always came with me. She kept the crew fed and put everything in order so we were always ready to go. Heading out to the Gulf Stream before dawn, Betty supplied breakfasts you've never seen before on a fishing boat. Our guys loved it, and they even reached a point where they didn't want to go fishing without her.

"If she doesn't come, we're not going to catch any fish … she's our good luck charm."

She could also make an equally good lunch in a six-foot sea while the boat rolled and pitched around. It was amazing to behold what she could do.

There were other great memories, like the time we won runner-up in 1994 at the Blue Marlin Association of Camachee Cove in St. Augustine. Harry Graves

Eric's graduation from the University of Florida

Tom's wedding, March 2005

won it that year, and I remember standing up at the microphone, accepting the second-place trophy.

"Harry," I said, "you're going to be up here handing *me* a green jacket next year."

That's exactly what happened. I won their prize-winning green jacket, and went on to win another one in 1996. The following year when someone else won it, the announcer said, "We finally got the trophy back to the working people of the club." Very funny.

The association was kind of odd about things ... it took two years for them to inscribe our name on the trophy, which shows how jealous they were. It's my opinion they didn't care for an outsider winning so many of their tournaments. We probably won ten tournaments in that club, but since we didn't keep the boat at Camachee Cove (where the association is based), we were considered outsiders.

A Word about Creativity

At Miller, we promote the culture of creativity. For example, we encourage all our estimators to come up with new ideas or ways to perform normal tasks. Right now our primary goal is figuring out how to prefabricate jobs, so one of our newer divisions is tasked with doing just that. Though it was hard to get going, we've slowly established that as a priority and have introduced more and more prefab systems for our projects. All this was brought about by the company's creative thinkers who came up with new ideas on how to get jobs done quicker, better, and with less labor. Labor can be our greatest savings or our greatest cost, so we rely on the creative thinking of our employees to develop ways to get the job done more efficiently. And by the way, anyone can be creative as long as they don't fear being wrong ... so speak up!

FIVE
Business and Politics

Then There Were the Unions

Earlier in my career I had to deal with some very colorful individuals in organized labor. My role was to negotiate agreements with them … the agents in the International Brotherhood of Electrical Workers, or IBEW. These were elected representatives whose full-time job was to make sure their members received their fair share, and otherwise do what was right for them. In all fairness most of them were, and are, good people who have to do what they think best for their members. Unfortunately some of them were real characters who went about their jobs in unseemly ways—outright thugs who used intimidation as their main tool for negotiating.

Dealing with agents often involved working out of town in different jurisdictions. I recall one encounter in Augusta with a very peculiar one. We sat down in his office and before we began talking, he pulled out a handgun and placed it on the desk. Being familiar with firearms, I casually reached over and picked it up.

"That's some kind of pistol you have here," I said. "A Colt .38 Special, right?"

"Uh-huh," he said proudly.

I opened the magazine and dropped the six bullets into my hand. Then I replaced the magazine and placed the gun on the table.

"These are nice," I said, rolling the bullets around in my hand, "125 grain Montana Gold, right?"

He raised his eyebrows. We went on to discuss the merits of his gun and the bullets, until I slid them into my pocket.

"Now, do you want to play with guns or do you want to negotiate?" I asked.

With Mayor Ed Austin and Judge John Moore

With Elizabeth Dole - late 1980s

With Mayor John Delaney at the Florida-Georgia game

Boy Scout Tour - 1986

With Mayor Ed Austin at our grand opening - 1991

With Jeb Bush - 1998

Betty and me with Mayor Jake Godbold

Bob Higgins, me, President Gerald Ford, and Bob Colgan

"Give me my bullets," he said, laughing.

"I will," I said, "after we negotiate."

You had to understand those guys. They were pretty tough, so you had to be just as tough to be effective as a leader ... at least with the union guys. They had no finesse in their negotiating style, so sometimes they tended to bull-head their way through. I have to say at times it was fun, and I learned a lot about people when dealing with those characters.

At the end of the meeting, he got his bullets and we had a deal.

***

There was one time in Panama City when I had to negotiate with an Italian union business agent named Albert Giordano. He was as tough as his name sounded. The meeting came about after I sent one of our superintendents for the job we were doing. But the local guys didn't like the way he ran things, so they actually shot at him in his motel room one evening. They almost ran him off the road on his way to work one morning. That was enough for our superintendent, so he ran back to Jacksonville and quit. I had to send someone else over there. My meeting with Albert went surprisingly well, and we had few problems after that.

Later I had to deal with one of the tougher situations I've encountered. It had to do with the business agent at the St. Joe Paper Company. St. Joe was owned by Florida East Coast Railway, and Mr. Dan Franklin, a good friend of ours, ran the operation. Though the plant was basically nonunion, we still had a union contractor working there. The result was as expected ... pairing union with nonunion was a bad marriage from the start. It took a lot of delicate negotiations to smooth things out, but we did. Incidentally, Mr. Dandelake negotiated that job for Miller before I took over. It was a "cheap fee" job and we should have been reimbursed for our payrolls every week. That didn't happen and the company was constantly four or five weeks behind in payments, which caused major cash flow problems. So I sat down with Dan Franklin.

"Look," I said, "we have a problem on this job. In a way you're like our banker and we're running way behind on our payments. As it is, we have to go to you to borrow money to meet our payrolls because you're holding them up."

His response was typical Dan Franklin.

"Son, that sounds like a good deal for me!"

The next week, however, he straightened it out and we were paid every week thereafter.

Once, in North Carolina I dealt with a situation at another paper mill ... union business managers again. I flew up to meet with an agent from the local union responsible for supplying labor for our job. His name was Smithson, and I knew he was an agent the moment he pulled up at the curb. Fresh from the set of *Urban Cowboy*, he wore a broad, white hat and drove a sports car—a Thunderbird, I believe—with a rack of steer horns mounted on his hood behind the chrome ornament. When I climbed in I couldn't miss the brown leather holster next to the center console. It had a gun in it. I suppose that, plus his CB radio, were there to show me how tough he was. In fact he *was* tough.

The problem was the pay scale. It was very low there and we had to pay more in order to get labor to come to the plant. While that didn't sit too well with the local contractors, we didn't worry about it much. The way it was, we might have put 200 or 300 people on a job so we took it upon ourselves to pay more than the regular scale. I worked it out with Smithson, and with our supplement we were able to get a couple hundred people on the job. He was a hard-hitting negotiator, but after working through the real issues we got along pretty well. At least this time the gun stayed in the holster.

Later in my career I dealt with the union through the IBEW president, Charlie Pillard. He was a tough old bird, but I had a way of getting under his skin one way or another. For some reason Charlie liked me ... or maybe he respected me because I was always honest with him. I didn't try to pull the wool over his eyes and I told him the way things were, saying if he would work with us, we'd work with him. Eventually, though, we would come up with an agreement benefitting both of our organizations. Maybe my stark honesty is what he needed.

During my terms as vice president and president of NECA, Charlie and the IBEW worked with us to bring about many changes benefitting industry. Although a lot of our members didn't think so at the time, they came to realize the changes we made benefitted them as individual contractors as well as the overall industry.

Strike!

Back when I was a foreman at the St. Regis paper mill I managed a team of men in the caustic area—a designated space in the plant where they recycled chemicals. When we neared the end of the job our final task was to tape the high-voltage buses in the switchgear. They were shipped to us un-taped but St. Regis wanted them insulated, so we sat in a cubicle placing electrical insulating tape on all the buses. One particular journeyman I assigned to the job was an elderly gentleman who took advantage of the situation. He managed to stay on the job for about a week, when actually a half day should have done it.

I kept urging him to finish up, but he stayed in his go-slow mode. When I kept on him, he grew belligerent. That happened several times, until I finally told him he was having a negative effect on the rest of the crew. He kept running his mouth and going on about how he was such a good friend of Jim Dandelake. Apparently he and Dandelake ran a radio station together some time in the past, and the journeyman thought it gave him license to do what he wanted. He also thought there was nothing I could do about it.

I finally had enough and met with our superintendent, Ed Hull, to explain the situation. I told him I couldn't work with such a man on my crew anymore. Ed told me he couldn't transfer him and said I should go ahead and fire him. I told him I'd probably get in trouble if I did, but I said I'd fire him if that's what it would take to fix the situation. So I did.

The next morning about 150 of our workers showed up at the site. They didn't go to work, though, but instead milled around near the fence by the railroad tracks, on strike. Several of them shouted, "One of our brothers was fired for no reason." They stayed a while longer and after a while went home without doing any work.

The next morning the workers showed up again, along with the local union business agent and Mr. Dandelake. The group gathered in front of the facility for a public meeting and the agent stepped onto a box to speak. He told them they had to get back to work and let the grievance process run its course. He also said the local union would probably file charges against Buck Autrey on behalf of one of the brother members, and so forth.

Then Jim Dandelake stepped onto a box. With a powerful voice he told them he was a friend of the journeyman who was fired and told them he still expected every man to do his work. He also said he respected Buck Autrey for looking out for the customer and not allowing someone to sandbag the job. Then he said he was 100 percent behind me and if they wanted to stay on strike, they were free to do so. If they did, however, he said he would terminate every one of them and start rehiring the next morning.

That was the end of the strike. It was also a big gesture of support for me from Mr. Dandelake. It would have been much easier to cave in to the union demands and get me off the job, but he refused. Instead, he took a chance and stood up for what was right. It was a valuable lesson I'll never forget … when people do the job right, you don't turn on them. You support them all the way.

The following week I was promoted to general foreman in charge of half of the crew on the job, with several other foremen working under me.

The Electrical Contracting Association

Miller Electric has always had a strong relationship with industry partners. Among the two most important are the International Brotherhood of Electrical Workers (IBEW)—Miller is a union shop—and the National Electrical Contractors Association (NECA), the largest and most powerful association in our industry. Henry Miller was the governor of our local NECA chapter from the 1940s, up until the time he died in his automobile accident. After that, Jim Dandelake, my predecessor, became involved with the local NECA chapter. But before I go on, let me give some background on our company's role with associations.

NECA started soon after Thomas Edison introduced the electric light. Almost from the moment he dazzled the public with his invention, electrical contractors rose from the ranks of Edison's Electric Illuminating Companies, springing up around the country like kudzu. As far as history records, the first contracting business opened shop in New York City in 1882. Not long after, hundreds of contracting companies popped up in every major city in the nation, and of course that marked the beginning of our industry's problems. As electricity took over it became apparent there were too many loose cannons in play and only teamwork among contractors would assure success for everyone.

As soon as NECA became official it began to tackle problems such as lack of uniformity in manufacturing specifications, inconsistency in laws and regulations governing electrical construction, lack of standards, a system of uniform training, and as always ... fair labor relations practices. The core problems they faced at the time remain with us today, although in greater numbers and with greater complexity. So while the association took an early, important role in the development of our industry, new problems continued to crop up. By the time I came along there was a lot of work to be done.

I started out by attending our local NECA meetings in the late '50s, when I taught in the apprenticeship school. I was interested in both teaching and attending the local NECA meetings for two reasons: to learn more and to get to know the people in the association. After I was promoted to senior estimator for the company in the '60s, I became even more involved in NECA. I cut my teeth serving on the local NECA Labor-Management Committee and the Apprenticeship Training Committee.

In 1970 I became governor of the local chapter, a position giving me an even broader exposure to the industry. As governor I sat in on the board meetings of the local chapter and served as liaison between the chapter and the national association. I also attended the national convention and was part of the governors' meeting there. I became fairly popular with the other presidents and governors in Florida and throughout the Southeast. That helped me get to know many of our competitors ... extremely valuable information, as you can imagine.

In 1973 I decided to run for the higher NECA office of vice president of District Three, the Southeastern District of NECA. They elected me vice president in the fall, and I served in that capacity for six years, from 1973 until 1980. It was an even better position, with greater visibility, since I represented NECA member contractors in the southeastern United States. In 1979 during my last year as vice president, my group decided I should be the next NECA president. Our Executive Committee nominated me for the position and I was elected. I took over in January 1980 and served as national president of the NECA for six years. That's when things really got busy.

As NECA president I also chaired the NECA Research Committee, the Manpower Development Committee, co-chaired the Council on Industrial Relations, and co-chaired the National Joint Apprenticeship Committee. Of course this was in addition to being the full-time president of Miller Electric.

The role of NECA president normally lasts four years, with the initial two-year term followed by a reelection for two more years. By happenstance, at the end of my fourth year the incoming president nominated by the Executive Committee decided to go "open shop" with his company—to go nonunion. Of course the national trade association, with its close affiliation with the IBEW union, could not afford to have a nonunion member as president. As a result the Executive Committee called a special election meeting (of which I was a member) and we withdrew the nomination of that individual. We couldn't agree on anyone else to take the job, so the committee persuaded me to stay on for another two years, until they got their act together and decided who would be the next president. I agreed to stay on.

We couldn't know it at the time, but their decision cleared the way for key improvements for the association and the industry nationwide.

A Seminal Event

An amazing thing happened at the national convention when I presided over my first board of governors meeting. The board made a surprise move and actually voted down a motion to approve the extension of the National Electrical Benefit Fund. Their decision eliminated the pension fund and, in so doing, essentially did away with NECA's relationship with the IBEW! I'll never forget it. Why would they decide to cut off such a valuable resource for our industry?

Obviously I couldn't let it happen, so I called an immediate recess to the meeting and decided to meet with several governors I thought would help fix the problem. These were friends I developed through the years, serving on the board of governors. I found as many as I could and we gathered in an empty conference room, where I stated my case.

"Gentlemen, we can't let this happen," I said. "This is going to destroy our relationship and everything we tried to build up with the IBEW ... all because

Ducks in a Row

we're having trouble with apprenticeship training issues." I added: "That will work out over time, but we can't stop everything because it will undo all we've done for the last eighty years."

I persuaded one governor after another to make a motion to reconsider the previous question ... in other words, to *reverse* their decision. During our discussion I also promised that Bob Higgins—the chief executive of NECA—would be more forthcoming with things related to the situation (Bob had a few communication issues with some of the members). I also noted he would retire during my term as president ... a prospect that would make the board's decision much easier.

We filed out of the room, reconvened the meeting, and the governors all lined up to give their speeches. When they finished, the motion for reconsideration of the decision was put forth, voted on, and accepted. The move reversed the board's decision they had made only two hours before, and the extension of the National Electrical Benefit Fund was approved after all. So, basically our fast re-vote was tantamount to reversing the direction of a fast-moving freight train on a dime.

That afternoon's drama was a major turning point in the association for two reasons: it was a test of my leadership ability and it was a test of NECA's relationship with the IBEW, where the latter had to agree to some action on their part to get things done. Considering the leverage unions often have, the resolution meant we stated our position well and exercised a greater degree of power.

There was still more background to the event. The night before the board meeting I met with the president of the IBEW and told him what would happen if he didn't comply with our position. He was reluctant ... NECA had recently received a court injunction preventing NECA from collecting industry funds for various programs. It wasn't an actual injunction, though; rather it was voluntary, so we needed the president of IBEW to sign an agreement stating it would be okay to collect the payments unless and until the courts found it null and void. He refused to sign the agreement at the time. But that night I told him if he refused to sign, it would mean the entire national agreement may go out the window at the board of governors meeting. Such a move would include payments into the National Electrical Benefit Fund, which would be a real sore spot for the IBEW. That was the only weapon we had.

Even though he finally signed the agreement, the governors were still upset over the industry fund issue as well as a few other clauses. The fact that he didn't enforce the apprenticeship clause was the biggest thing. Additionally, failing to enforce the apprenticeship clause guaranteeing one apprentice to every three journeymen throughout the industry upset the board of governors more than anything else. That's why the board voted to do away with the national agreement which would, in effect, cancel the industry fund, cancel the national electrical benefit plan payments, cancel everything … and revert them to the 1 percent instead of 3 percent member payments. We had to get it revised during the crucial board of governors meeting, which is the reason the IBEW chief had to agree to what we wanted done after the board of governors voted it down.

So much for politics.

As a result of the vote I gained the respect of a lot of governors and others in the industry. Although I was up and above board, I was still accused of being kind of autocratic when it came to gaining the vote, but we had to do whatever it took to get it passed. Later on, an inspired vice president who was there drew a portrait of me having "all my ducks in a row." He showed it to the members at the next meeting.

***

Another key event during my tenure was that situation with Bob Higgins, our top NECA executive. He had been with the association a long time and became a root cause of a lot of dissention among the members. As a result of our prodding, he ultimately decided it was best to retire. After he resigned I interviewed several potential NECA CEOs, and when we selected the right one I lobbied the committee to hire him. His name is John Grau and he became our new executive in 1985. John has done an excellent job in the past thirty years, and I've asked him to talk about NECA later in the book.

Changing the Industry

Another milestone during my time as president was our move to eliminate labor-related services to nonunion contractor members. NECA had about 400 *nonunion* members at the time of the proposal, so our resolution of course would most likely eliminate all of them. It was tough to do from a relationship standpoint since their dues were a minor part of the association's support. The move was also necessary because of situations like the contractor receiving the nomination for president after his company went nonunion—the reason I stayed on as president for a third term. Yet another reason was our need to guarantee IBEW that we would be a 100 percent supporting trade association for them. Finally, the resolution would assure us future concessions from IBEW. We passed the resolution and in return received full cooperation from IBEW nationwide. In the short term we lost many members, but far more positive results would materialize in the years to come.

The resolution helped us hold our own through better working conditions, helped get rid of archaic restrictions in local union agreements, and attained a better ratio of lower-paid workers such as apprentices and pre-apprentices. We also established a pre-apprentice program where young people could come in and work at a lower wage. The ratios were adjusted to one-to-one instead of one-to-five or one-to-three. We could actually have a lower-paid worker right there with a journeyman, which considerably lowered our crew cost to where we competed with nonunion wages. It was a huge development and it came out of what we call our National Agreement negotiated in the '70s, for which we had to go through all kinds of lawsuits and negotiations to finally get established.

In order for us to get those concessions, we had to commit to the IBEW, to support them as an organization in exchange for their being a full partner with our association. The arrangement has proven to be beneficial to contractors nationwide. We have better ratios, better working conditions, and almost unlimited portability of our workers practically anywhere in the country. All of these conditions came out of the critical partnership formed in the '80s and early '90s. That agreement between labor and our association was the industry equivalent to the formation of NAFTA, where mutually beneficial business channels opened wide the way to prosperity.

The Aftermath

NECA had close to 6,000 member companies at the time I became president. But after the nonunion contractors and some others pulled out, we were left with around 4,000 members ... fully one-third of our membership. Later, during the recession in the '90s, many older companies either left the association, died off, or retired. At the same time a lot of the smaller companies who didn't think they could survive as union shops pulled out and went nonunion. Still others—and there were many—went bankrupt. We didn't lose any of the large companies, however, and if revenues dipped in the short run we eventually regained our income footing quite well.

To serve as president of NECA for six years was significant in itself. As it would turn out, the timing of my presidency would have a lasting impact on the association and the entire North American electrical contracting industry. The decisions we made at the time were good ones and they definitely helped both NECA and the IBEW. We have retained our membership levels and secured our portion in the marketplace. The nonunions still do the majority of the electrical work in the country, but the IBEW holds its own in the industry while other unions have steadily gone downhill.

Leading the board of governors to adopt resolutions allowing these things to come about were the highlights of my presidency. It was an amazing convergence of timing and position: the Justice Department investigation (which I'll address later on), the transformation in the industry, the economy, and leadership skills ... all converged as a critical situation to bring about much-needed improvement. Many key decisions had to be made, and although iffy at times, thank Goodness we made the right ones.

Beyond the Presidency

As a past president of NECA, I automatically became chairman of the Academy of Electrical Contracting ... a group of officers in the trade association and other elected individuals. The academy was a think tank convening a couple of times a year to make decisions and write papers on subject matters benefitting the electrical industry. As such, many of the policies of NECA adopted by the board of governors came out of the Academy. For me the academy was yet another platform to address the points of concern and otherwise explore topics affecting the electrical industry in general.

During my term as Academy chairman from 1986 until 1989, we brought up the idea of establishing a foundation to study different aspects of the industry, like how labor affects variables such as overtime and productivity. The proposal was dormant for a couple years, until the board of governors finally adopted a resolution authorizing the association to establish a foundation. That's when they asked me to commit $100,000 to spearhead its formation.

We did, and Miller Electric became the first contributing member of what we called the "Foundation of Electrical Contractors." The name was later changed to "Electri 21" or "Electri International," in deference to the twenty-first century. Our contribution set the benchmark for other contractors and provided the leadership to build a permanent endowment to fund the foundation's programs.

After Miller's commitment, they asked me to chair the foundation for the first year to help them get going and, of course, bring more contractors on board. I said I would, and within a year we raised several million dollars and attracted several new contractors to the program. Within two years the foundation was up and running with several million dollars in reserve. Since then the foundation has initiated studies and published many papers ... all to the benefit of the industry.

***

There is a postscript to those busy, productive years with NECA. I'm very grateful to Ed Witt who was behind me, practically running the whole company in my absence. My daughter, Susan, and son, Ron, were also a tremendous help, but Ed was the principal force behind me. We communicated daily, no matter where I was. Freeing me up like that helped me do what I had to do. And yes, it was good to be so involved ... the process built a lot of character in me as I passed through the fire of desperately needed change in the industry.

On Politics

As my career developed and I became more involved in leading elements of the electrical industry association, I found myself playing the role of a politician at times. Since that was the case, I actually considered going into politics one day. My wife put a quick stop to it.

"You can be president of NECA," she said, "and you can do anything in the business you want, but don't you ever dare to get into politics!"

That was that. However, if I wanted to, there were several opportunities to get involved in local politics. I probably could have run for mayor and would have made a good showing due to my friendship with several mayors in Jacksonville. And because I was in business on a rather large scale, I was politically involved anyway. It came with the position and I couldn't avoid it. After all, politics and business go hand in hand and it helps protect our interests in many ways. Nevertheless, I kept business at the forefront and remained behind the scenes as much as possible.

Deep inside, however, I never wanted to be in the political scene ... certainly not after they passed the Florida Sunshine Law. If I entered politics in a big way, the law would require me to expose everything I've done in my life. I simply did not want to do it. My assets would have to be transparent: my income, debt, expenses, net worth ... everything. That's "government in the sunshine," where the public has a right to know everything. Although I have nothing to hide, I've always been a private person where my private life and business are concerned. In fact, the law made it difficult for businesspeople to get into politics, period.

Betty was smart. She didn't think political life was a good place for me, and I agree with her now. But the tendency is natural because if a person thinks he's a good leader and has been successful in business, they'll often aspire to go into politics. You see it all the time. Businesspeople such as Donald Trump—although I don't think he would stand a chance of being president—got the idea to go into politics. Then there was Mitt Romney, another successful businessman I hoped would be elected. He had the right message, even if it was the wrong time for him. No, I never had a strong desire to be a politician ... but I certainly wanted to influence them.

Written in Ink: Sandy Livingston

I started with Miller Electric in August 1993, after working a couple of different businesses from my home when my children were young. Through those years I dabbled in real estate and had an engraving business. Then my brother, Ron, offered to send me to school to learn 2-D drafting on the computer. He saw a need for this position in the company and offered it to me. I've always enjoyed drafting, drawing up house plans, and playing around with computers anyway, so it seemed a perfect fit. I took him up on it and have been here ever since.

The first thing that comes to mind about my father is his good work ethic. He's a hardworking man and he's never afraid to jump into a project, whether fixing a sprinkler system or some kind of repair on his boat. He's a hard worker and he appreciates people who work hard. Dad is also hard to please at times, but when you please him he shows his appreciation. He also has a big heart. He adores his family, especially his great-grandchildren, and is a devoted husband and father. Although our roles here don't cross much, what surprises me most is my father always seems interested, even impressed, with the work I do … which gives me a wonderful feeling of accomplishment.

In my opinion my father made Miller Electric what it is today. He has always thought of his employees first and insured they were taken care of, especially those who worked away from home for long periods of time. His day-to-day involvement in the company, his strong role with NECA, his constant desire to learn, and his focus on continuing education in the business, have helped develop what we are today. And although he's partially retired, his continued presence in the office most days shows everyone he's still interested and involved. He has created an environment inspiring those of us working here to be the best at our jobs.

My dad's a straightforward guy and, although not a practical joker, he can be funny when it comes to his social media skills. He's always trying to stay relevant by doing Facebook and texting, using all the abbreviations and so on. The other day someone got a text from my dad that said "LOL." We thought it was hilarious. He enjoys learning about a new app on his phone or some software on his PC. It's important to him to stay relevant.

If my father wasn't at his desk, he would be on his boat, of course. He'd be cruising or fishing, or if at home he'd be at the kitchen table doing his daily crossword puzzle. He does them in ink, by the way, taking pride in knowing he has the right answer. He loves his crossword puzzles and you'd better be on your toes because he expects you to know the answers to the crazy crossword language!

As for advice from Dad, I remember what he told me as a teenager: "When you're in a discussion with somebody and you want to say something, run it through your mind three times. If it still sounds good, go ahead and say it." I probably have not always followed his advice to the letter—just ask my husband! And if there was anything about my father I'd change, it would be that he'd have good knees and normal hearing.

My family has been involved with Miller Electric for a long time—forty years for my husband, who retired five years ago. He traveled throughout the Southeast for his job, which enabled us to meet some fantastic people. My son-in-law worked for Miller throughout his years in high school and college and is now one of our VPs. My daughter worked for Miller while going to college pursuing her MBA, and left to have our first grandchild. My son, after graduating from the University of Florida, wanted to come up through the business the way his grandfather did. He enrolled in the electrical apprenticeship program, graduated with the Top Apprentice Award, and is now one of our assistant project managers. He is a fourth-generation electrician, and I'm proud of their involvement and accomplishments.

Miller Electric has been a part of my life as long as I can remember. The more time Dad spent in the business and as his children and grandchildren grew, the more he realized the need to bring in family to continue to grow the business. That's not to say he was tired of the business ... not at all. He saw needs the family members could help fulfill while maintaining the family culture.

Dad likes to refer to the company as a family business even though it's grown so much through the years! It's huge, but he still likes it to feel like a family; it's important to him that his employees have the same feeling.

Sandy Livingston

Manager – BIM, 3-D Design

A Word about Choosing Good People

Part of good leadership is the ability to understand people. If you understand people and their character, you have a better chance of selecting the right person for the job. I think that's one of my best attributes: understanding people well enough to place them in the right position. I have mentioned before the best and most loyal people are those we promote from the field. The absolute best way to build a company is to choose those who have worked for the company from day one ... those who have both the ability and the desire to do better. An important thing about good people is their strong desire to be better tomorrow than they are today. That has always been my motivation, to be better than I was the day before or the year before.

When we bring people in from the field, we don't throw them into the pot, so to speak. First they work with other estimators to absorb the team atmosphere, and after a long period of time they might become a manager. Even if they're a good field manager when they come in, it doesn't make them a great project manager. We use the term "projects" in the plural since a person has to handle multiple jobs and gain the ability to multitask. After they've gained some ability and have a good track record, they can consider themselves successful. The contracting business can be iffy, often without a long-range plan. That means we have to work hard with what we have right now, which leads to picking the right person. They need to be flexible and work in our ever-changing environment. Eventually, through their dedication and hard work they might become a project manager. And when that happens they'll be well-rewarded.

SIX
Antitrust

The Grand Jury

As noted earlier, my tenure as NECA president was traumatic at times. That was certainly the case when the entire electrical contracting industry came under investigation by the Justice Department's Antitrust Division. They implicated several large contractors suspected of price-fixing, based on complaints from cities throughout the nation. One specific job in Virginia came to light after a dissatisfied customer notified the authorities, causing the Justice Department to look into it. One thing led to another and it snowballed into a full-fledged investigation of the top five contractors in the nation: Fischbach & Moore, E. C. Ernst, Foley Electric Company, L.K. Comstock & Company, and Lord Electric Company. These were large unionized companies and members of NECA. Several smaller contractors were also pulled into the mess.

To address the perceived widespread problem, a grand jury convened in Washington, DC, at the same time as ten other grand juries convened throughout the United States. As they proceeded, they decided the most probable cause of the whole thing must be the contractors' association, NECA. Naturally, as NECA's president, I became deeply involved.

I had to go before the federal grand jury several times to convince them I was not "The Godfather" (a term I often heard) of the crimes committed in the country. Of course they didn't buy it, so I had to stay involved in the process over the next four years.

I met with the Justice Department several times to try to persuade them to close down the investigation and let us do our own in-house policing. We said we'd discipline our members and launch a nationwide public relations program to prevent our members from getting ensnared in the process. Of course they wouldn't shut the investigation down, and needless to say I didn't feel good about it.

Ronald Reagan was president and he was being accused of rolling back much of the legislation against white-collar crimes. So there was slim chance of getting any help from the administration. The head of the Justice Department at the time was Judge Bork … the same Bork who was later nominated for the Supreme Court and subsequently "Borked out" by the Democratic Congress. Among his other duties, Judge Bork served as head of the Antitrust Division, so I arranged to meet one-on-one with him to ask for help. I told him what the association planned to do, and laid out our plan of action. We had a good discussion and he listened carefully, but he still couldn't shut down the investigation.

Ultimately the Justice Department indicted several companies for price-fixing contract bids, alleging they did so starting in the '60s and on through the '70s, for all kinds of projects. They accused them of targeting big government contracts such as nuclear power plant facilities.

Here is the wording of the original **U.S. Court of Appeals for the Third Circuit – 750 F.2d 1183 (3d Cir. 1985). They were very specific with their accusations.**

"For approximately seven years, from 1974 to 1981, representatives of several electrical contracting companies met at the Duquesne Club in Pittsburgh. The purpose of these meetings was to allocate electrical construction projects, by bid rigging, at the Western Pennsylvania Works (Works) of United States Steel (USS). Whenever one of the companies desired to rig a bid, that company's representative would telephone the other contractors and arrange a meeting. After deciding among themselves which firm would receive the contract, that firm's representative would contact the other contractors and would tell them what bid to submit, ensuring his firm's bid would be the lowest. Some of the firms kept track of the allocations to ensure each contracting company received its fair share of work.

As a result of investigating this bid rigging scheme, the government accused a number of the electrical contractors and their employees of antitrust violations. The indictment charged them with violating Sec. 1 of the Sherman Act by conspiring to allocate electrical construction projects at the Western Pennsylvania Works, to fix the prices at which those projects were bid, and to submit noncompetitive, collusive, and rigged bids on those projects."

Later, the document went on to say:

"The government proved the defendants met, over a period of several years, at the Duquesne Club for the sole purpose of rigging bids and allocating contracts. The government introduced evidence to prove the participants knew the purpose of the club meetings and they met as a group, thus negating the argument that separate groups of the defendants were conspiring together, with the same goal, but with their actions unconnected. More important, the government showed some of the defendants kept records of which company received which job, ensuring each firm would receive its fair allotment of the contracts. This recordkeeping indicated a single conspiracy continued over time, and not separate conspiracies involving only certain contracts."

It is worthy to note that although I was the president of NECA during these proceedings, Miller Electric Company was never implicated by the grand jury for any wrongdoing throughout the entire process. It's also worthy to consider why these companies collaborated in price-fixing in the first place—and later went out of business. They simply didn't know how to compete. Once the Justice Department cracked down on their backroom deals, they didn't know how to effectively run a business.

When It Rains …

Even though Judge Bork and the antitrust people denied our request, our association proceeded with the public relations plan. The board of governors adopted a resolution designed to highlight the ills of price-fixing and the consequences that offenders could expect from the Justice Department. While we were praised by the Justice Department for our actions, in the end it actually attracted a few lawsuits against NECA. The reason is around that time we were also in the middle of a lawsuit against us by the National Constructors Association, or NCA. Their lawsuit wasn't about price-fixing, but rather over an agreement we (NECA) negotiated with the IBEW. The origin of the lawsuit goes back to the '70s, when we passed the agreement that included several clauses NCA objected to, specifically the "Industry Fund." To recap, the fund required employers of electrical workers from the IBEW to contribute 1 percent of their salary to the national office for the operation of the industry fund. The fund was supposed to take care of all supervisory training and part of the apprenticeship training ... in other words, for training purposes only. The bigger contractors had a real problem paying 1 percent of their payroll to the association, so they filed a lawsuit to stop it, as a restraint of trade.

The initial court decision was not in our favor, and declared NECA in violation of "restraint of trade" laws and therefore the action was deemed illegal. So in addition to my trips to DC and other places for the grand jury situation, I was often on a plane to New York and San Francisco to try to get people together on a settlement and get the lawsuits behind us.

***

Of course there was a lot of publicity about the antitrust fiasco. Articles frequently appeared in the *Washington Post* about large Washington, DC, contractors such as Foley Electric, or the *New York Times* about Fischbach & Moore—one of the largest publicly-traded electrical contractors of the day—and the *Wall Street Journal*, among others. While the companies accused of antitrust created a field day for the newspapers, NECA was largely spared the publicity. It might be due to our visit to Judge Bork, and by our many testimonies in front of the grand jury.

Our local Jacksonville paper carried a story about Paxson Electric, the company Wesley Paxson formed during our survival days at Miller. Paxson was one of the organizations investigated by the grand jury and they fought the allegations all the way to the appeals court. The ultimate decision was detailed in United States v. Fischbach and Moore Inc. – 776 F.2d 839:

> "Paxson's declarations were to the effect the project was not the subject of rigged or agreed bids. Paxson was convicted in the Northern District of Georgia, as were his company (Paxson Electric), four other companies, and an executive of one of those companies for a violation of the Sherman Act, and two counts of mail fraud in connection with the rigging of that project."

At the end of their long discussion they concluded:

> "For the reasons set forth above, we find no reversible error and therefore affirm the judgment of the District Court."

Wesley Paxson served a couple years in prison. Several years later after he got out, Wesley called and asked if we could talk. We met at a downtown coffee shop and he said he wanted me to buy his business. I already knew his company had been on the market and he only asked me after he was unsuccessful selling it to someone else. I can't speculate on his reasons, but maybe the antitrust ordeal sort of broke his spirit for the business. The whole thing cost him a lot of money—probably in excess of $3 million counting his legal fees, fines, and everything else. That was a lot of money then, and would have put most companies out of business.

He made his pitch and then told me how much he wanted for his company. I said I didn't think we could afford it but would make a counteroffer, which I did. He said it wasn't enough, so I told him I was sorry and we couldn't do business. Two months later in January 1992, he called and said he wanted to meet one more time to see if we could work it out.

That time I sweetened the pot somewhat and he agreed to most of my terms. We shook hands and formed a joint venture on the work he had in process, which was substantial at the time. We also assumed leadership and responsibility for many of his jobs and over the next couple of years—with the help of Paxson's additional business—we doubled Miller's volume. That moved us up to the level of a $90 million company. We also made more profit because, well, we no longer competed with Paxson.

A Word about Intuition

I never relied much on intuition. I'm a more logical type person and if something doesn't seem logical to me, it probably isn't. Likewise if it seems too good to be true, it probably is. Living by those guidelines means there isn't much room for intuition in my decision-making. I believe if someone relies on intuition too much, they're actually gambling. And there's enough risk in this business without having to gamble.

In my opinion the only time intuition comes into play in management is during a job interview. I can't pull all the facts from their background or have a sense of what's going on in their mind when they're right in front of me. Then I use my gut feeling and think: *I'm not getting the whole story, but I think he has what we need. He has potential.* So when it comes to intuition, I guess I take a little chance with new people. In the contracting business, though, we have to rely on facts and figures more than anything else.

No Job Too Small: Daniel Brown

My grandfather is a highly respected man who is confident in himself and everything he says. At the same time he is not prideful or boastful in any way, yet there's a certain confidence in knowing what he's doing is right. I think a lot of people respect him for that as well as for his intellect, knowledge of the industry, character, honesty, and the way he conducts himself.

My first memories of my grandfather are going with him to the Gator football games and family gatherings. But he traveled a lot in the early and mid-'80s, so I don't remember much about those years. One of the more detailed memories happened when I was around ten. It was my first fishing trip when he took me and my brother, Henry, along. We

---

tried several times before that, but the trips were always canceled due to bad weather. Every time that happened, I was devastated. When we finally got to go fishing, it changed my life and opened up a whole new avenue for me. My grandfather said it changed his life too. During the next fifteen years I was very much involved in fishing, spending most weekends with my grandparents on their boat. I've continued to fish up until today.

Papa has helped me in my work and life through fishing ... not the sport itself, but the way he went about it. We spent a lot of time fishing tournaments in the Bahamas and St. Augustine among many big boats, most with hired captains and crews. At the end of the day the owners would get off the boat and grab a cocktail somewhere, while the crew cleaned the fish and the boat. If there was any work to be done in the engine room or below decks, the hired hands did it. Well, Grandpa wasn't that way. If something needed to be fixed, he did it himself and was proud to do so. He didn't clean a lot of fish and he didn't clean the boat that often, yet any other work needing attention he did himself. So he was almost, could I say, looked down-upon in the billfish tournament circuit. That didn't matter to him because that's the way he was and that's the way he would always be. The crew and I greatly respected him for it.

Spending time fishing with my grandfather helped me learn about his character. He had all the money he needed and owned a big boat, yet he wasn't afraid to get down and get his hands dirty. He showed me no matter how high one might be in life or career, no job is too small or menial. Observing him on the boat definitely had an impact on my work life, and my family life also.

My grandmother was also right there on the boat. She would never let the crew go hungry, which was something we kind of joked about. On other boats with hired hands, the owner would step on board with his cup of coffee and Danish while the crew did the work. But on our boat as we prepared for the day, we ate a plate full of grits, eggs, biscuits, bacon, and sausage! My grandmother was happy to do it ... sort of her way to brag on my grandfather's crew being well-fed and ready for the big day. My grandfather won the Northeast Florida Marlin Association's "Boat of the Year" award in 1995 and 1996. I was honored to be a part of the crew and elated to be there to see him win. It had been a long standing tradition for the winner to receive a green sport coat, and all the past winners would wear their green jackets to the awards dinner. I was once again honored more recently when my grandfather attended this year's awards dinner wearing his green jacket when he came up to the podium to present me with mine for being the 2015 Boat of the Year. It was an emotional night for me. It felt like a torch was being passed from him to me. I hope that I can keep it burning brightly and somehow use my love of fishing and family to have a positive impact on my boys' lives the way that Papa impacted mine.

Papa was always concerned about the younger generation in the company. He didn't want younger employees to come through the ranks faster than they should, or climb too high too fast. He wanted their hands to get dirty. He wanted them to study the drawings, count light fixtures, and otherwise spend time doing the job regardless of the position.

It's been three years since my grandfather officially set himself aside as chairman emeritus of the board of directors, yet he still comes in to work. He shows up several times a week if he's in town and walks around the various offices. When he does, I can still see a tremendous amount of respect in the eyes of the people who work here. I can tell they're happy he's still around. I think it brings comfort to them to know his watchful eye and strong, positive character are still here.

Daniel Brown
COO, Miller Electric

A Word about Keeping Good People

Obviously, profit sharing is a strong tie that binds. For people to stay on in the company, they have to know that in accordance with their own ability and effort, rewards will follow. And if they do even better, they'll be rewarded more through profit sharing and cash bonuses as well. Both of these are real motivators. Good

Landing on an aircraft carrier, 1992

people perform better when they know they're responsible for their own actions, and take the initiative. That's why we delegate complete authority to our people and hold them accountable, and they're accountable to the point it might cost them money if they make bad decisions. They know wrong decisions will affect their bonus at the end of the year. On the other hand, motivation by fear does not work. If a person is a professional and good at his job, he won't be afraid of losing it. He is confident if he makes a mistake he won't be fired. We've had people in this company lose more than $3 million on a single job and not get fired. The reason is some situations were not their fault because mistakes were made by others involved in the project too.

The most expensive thing we can do is fire a good employee. If it comes to that, a lot of money has already been spent to bring them up to speed and that person's replacement has to start from the beginning, which costs even more money. It has never, ever crossed my mind to fire someone unless they're unable to overcome their mistakes. If they constantly make the same mistake, we have to let them go.

Few people have left the company over the years. Some of those who did, eventually came back and asked for their old job back because the grass wasn't as green out there as they thought. In some cases, when we took them back they became excellent producers and leaders for the company. The company's culture of forgiving mistakes or misdirection—within reason—points to job security, which is a strong motivator for good employees to stay on. Add cash incentives and there's a strong case to stick with their job for a long time.

Employees can see that in my long career, as well as that of our current president David Long. David started in the field as an apprentice and cable splicer, worked his way up to become project manager, and eventually president. When he could, he went to school, took business courses, and established himself as the largest license holder we have in the company. David saw the opportunities in our company and is now one of the most dedicated employees we've ever had. And by the way, when it comes to licenses, my grandson, Daniel, is right behind David. Daniel takes licenses everywhere he's worked and as a result we have redundancy, the assurance of our ability to deliver services most anywhere.

SEVEN
The View from NECA

Perspectives by John Grau,
CEO of the National Electrical Contractors Association

A Gentle Twist of the Arm

In 1979 when the association elected Buck Autrey as president, I was a local chapter executive in Milwaukee. Aside from seeing him at various association functions, all I knew about Buck was his reputation as a bold, brash, dynamic head of a successful company. I also heard anyone who met Buck could tell he was a real leader, and based on his strong accent they knew he came from the South.

Five years later, in 1984, I finally met Buck in person when NECA began interviewing for a new CEO. The Hiring Committee had already interviewed several people for the job and had winnowed the number of candidates down to about a dozen. The final interviews took place at the convention in New Orleans.

It was a stifling southern day when I stood in front of the committee and in spite of the air conditioning, the room was hot and humid. I was one of the final candidates for the job, and Buck was in charge of asking the questions. It was a group situation. Four Selection Committee members had assembled behind a table while the rest of the Executive Committee sat in the room to observe. I introduced myself, answered their questions, and left the room. It would be a while until I'd hear the results.

That evening our Political Action Committee held a fundraiser on a Mississippi riverboat. The committee held a promotional deal where everybody on board had to give—or solicit—a minimum of $500 toward the effort. Buck was there, of course, and after the program got rolling, he cornered me.

"John," he said, leaning close with one hand against the wall above my shoulder and a martini in his other, "I need you to commit to $500 for the PAC."

I'd just interviewed earlier in the day for the job and now the chair of the Hiring Committee asks me for money? Of course I couldn't say no, but at the time $500 was a lot of cash for me. Fortunately, Buck qualified his request.

"Well," he said, "you don't have to give it yourself … you can go to the people in your chapter and get them to give some money."

I don't know if that awkward exchange qualified as real dialogue with Buck, but it was the beginning of our relationship. I didn't see him again until much later, after an Executive Committee meeting on Saint Martin, an island in the Caribbean. Bob Higgins, my predecessor, called to tell me I had been selected as his replacement. After I digested the news, Bob told me I should go and spend time with Buck and the others at their meeting. Apparently no one planned for me to attend, because there were no other rooms available. So I wound up sleeping in an extra room in Buck and Betty's suite at the hotel … a good first introduction on a rather personal level. That night Buck told me he was one of my supporters.

Not long after I came on board, Buck, Bob Higgins, and I hit the meeting and chapter circuit throughout the country. As we traveled I got to know Buck a little better. Once, after an event in Orlando, we jumped on a plane and headed for another meeting in Monaco. We transferred planes in Atlanta, flew to Paris, and drove down the south of France to Monaco. Part of the trip involved a helicopter ride over the beach along the Riviera. Bob insisted he sit in the front of the helicopter because, as he said, he'd "seen things happen before and if anything went wrong he could jump in and take over." When Buck heard that, he looked at me and rolled his eyes as if to say, "If Higgins jumps into the pilot's seat, I'm jumping out and I'd advise you to come right behind me."

Buck knows how to have a good time and sometimes he can act like a kid … in a good way. Of course along with the responsibilities he had, his serious side was evident as well. Early on I noticed how Buck made good use of the mealtime at our meetings. His table always included Bob Higgins and a few others. The "others" were those he had decided to either reward or chastise. If someone did well and Buck invited him or her to his table, he regaled the privileged one with his legendary stories. If someone crossed the line the wrong way, he or she had to sit through an entire evening listening to Buck's lectures. He'd tell them they'd better shape up and do it right. When I watched those sessions, I was impressed with the strategic way he accomplished what he wanted.

The Right Man at the Right Time

It was a good thing Buck became president of NECA when he did. He came in during one of the most tumultuous times we've ever had in the association—at least in the years I've known, and I'm sure even before that. It was a time we experienced a sharp drop-off in union density … meaning there were a lot more nonunion shops entering the electrical contracting market. If we consider the statistics, the six years Buck served as president were the years the drop-off occurred. I don't know all of it, but I think the cause was the reaction of big companies to the unions' tactics. It seemed to have started with a business roundtable group who decided they'd had enough of the union and the way they organized jobs. I believe that group inadvertently helped the nonunion competition grow.

We also had older leaders in the unions who were recalcitrant in making changes. Charlie Pillard, the president of IBEW at the time, had been in the position many years and was a true old-line union leader. So along comes Buck Autrey … a young, aggressive guy of around forty, whose presence widened the age and philosophy gap between the two.

Buck wasn't afraid to walk in and meet with the people from the business roundtable. He was president of a big contracting company and dealt with the Savannah River project, among others, so he knew how to relate to big business. Buck was also from the South, where many contractors had experience dealing with nonunion competition. He could offer them advice and direction. Since Miller was also a successful company, he was able to stand up in front of the membership and challenge them.

"Yeah, there's a lot of nonunion competition out there," he'd say, "but my company has been able to deal with it. We're also being hurt by it, but there's a way to get around it. You don't have to abandon the union. You should work with them instead."

Nor was Buck afraid to speak up to the union leadership. It helped that he could talk boldly about his own company and the progress they made with both business and the local union. He gained respect for it which, if attempted by another leader in a different position, might not have been successful. I think he did a lot to help NECA's members bridge several gaps where there was a lot of dissension.

Buck was not afraid to push back on some of our larger member companies, either. Some of them complained about their high association dues, which were calculated on the size of the member's business. Buck stood up in the meeting and challenged them again. "Gee, I wish I could pay so much in dues … it would mean I've got more work and I'm making a lot more money."

Among the many trials of the era was a major antitrust investigation into some of our bigger members. The issue was all over the press for a long time, and Buck traveled to meeting after meeting throughout the country. He also met with the federal grand jury several times to alleviate the problem.

Another significant change at the time involved NECA's leadership. Bob Higgins had been the association's CEO for twenty years, and I think Buck—along with others—thought it was time for a change. Buck helped Bob decide it was time to retire and make room for new leadership. It became a peaceful transition borne of mutual understanding … which leads to another dimension of Buck: he was good at bringing about change *within* the system. Although there were many changes during his tenure, Buck didn't do it by throwing bombs. That was a far different approach from those in NECA who would destroy the association or the union if they weren't happy with the situation. Still, there were a few who thought Buck's approach to change was too radical. Yet, he persisted and worked the system without tearing it apart.

Not the Usual Background

Although Buck didn't have a formal education, he was a savvy businessman who got things done. He was a self-taught, self-made man and where he got his knowledge and wisdom is a mystery. When I talked with his son Ron or their president David

Long, they constantly quoted things Buck taught them about his philosophy of business and life. We also saw Buck's influence on the younger employees and even the Autrey grandkids. "Oh yeah, Grandpa said this or that." They respect him for his sage wisdom, and so do I. I know of many things Buck said that rang true and I know he ran his business using those truths as well. They were the kind of principles found in books with quotes by famous people. It seems Buck internalized many of those philosophies and principles. I don't know if he heard them somewhere, read them, or came up with them on his own.

One more thing about Buck. We think of him as being around forever, but he was the leader of NECA when he was in his forties … quite unusual in our industry. The other presidents had already worked in their jobs and served for a long time before becoming president. If you study the photos of the Executive Committee members, you'll see their average age is around sixty.

The Political Buck

Buck was always savvy in labor-management relations. He knew when to fight, when to back down, and knew he couldn't win every time. He looked at the objective and determined what he wanted to accomplish. So when it came to labor, Buck wasn't about winning; he was about getting the right results, which included happy workers doing a good job for the company. We have many people in our industry who get into a fight and only want to win without knowing what they want to accomplish. Buck could see the bigger picture and he didn't allow his ego to get in the way of his objectives. Yes, he had an ego, but that didn't block him from doing what was expedient and correct.

One example is when NECA and IBEW tried to work out a particularly sticky issue with the National Electrical Benefit Fund. It happened during a meeting with our board of governors and it came to a vote. The governors turned it down and their decision was definitely against the wishes of the leadership, including Buck. The vote took place right before lunch, so a motion was made to adjourn for the break. After they returned, I remember Buck saying, "Okay, let's vote again."

I don't know if the move was legal or not, but when the governors voted again, it passed! I don't think there has ever been something like that in our association, ever. People complained about it and of course the story spread: "Do you remember the time Buck called for that vote again?" And, "I don't know if he could do it legally, but he did!"

As I heard it, Buck worked the crowd during the lunch break. He pulled the right people aside and had a word with them. Later, Charlie Pillard, the head of the IBEW at the time, told me his version.

"Yeah, I arrived at your convention to give my talk the next day and when I pulled up, there was Buck waiting for me at the entrance. I asked him what was the matter and Buck told me they—we—had a problem. I asked him what he needed from me, and he told me."

Buck wrestled the change he needed from Charlie Pillard, and returned to the luncheon to twist more arms. He was a one-man show who got the resolution passed, and remains one of a few who could get away with it. People still talk about it today.

Seed Money

The Electrical Contracting Foundation was established toward the end of Buck's presidency. Although he didn't champion the project, his actions after he left office definitely got it off the ground. The idea of the foundation was to conduct research on behalf of the industry, and funds for the foundation would come from a separate endowment. We hired a fundraising consultant to work out a plan, and the idea was to find someone to spearhead the cause ... preferably an organization willing to be the first to contribute. The consultant worked with us to set the bar for giving, and the next step was finding a lead company to make the first donation and encourage others to follow. The first company we approached was Miller Electric ... maybe because Buck was the immediate past president, which made it a likely choice.

The fundraising consultant and I flew to Jacksonville and met with Buck. As I recall, the figure of $20,000 was a good start for contributions, so the three of us sat down to go over the consultant's gift options list. It turned out to be a short meeting.

Buck said, "Okay, I'll do the $100,000."

"What?" I said. "The whole $100,000?"

"Yeah," said Buck. "It's $20,000 per year times five years, so that's $100,000, right?"

After the meeting the consultant told me what happened.

"We set the fund up for $20,000 per year," he said, "and if you multiply it out, it's $100,000, so Buck pledged to do the whole thing at once. He believed in it, so he did it. He said anybody who wanted to be involved would have to commit to buy in at $100,000."

That's how our founding level of giving was established. Buck and I visited numerous meetings around the country where he'd say, "I gave $100,000 and you should give $100,000." Within a year we had commitments for $7 million, and a couple of years later we had pledges for $10 million. Today the foundation is alive and well-funded with an endowment between $15 million and $20 million, depending on the stock market.

We did something similar with our Political Action Committee start-up funding. We asked for an initial contribution of $500 and later upped the amount to $5,000. Again, it was Buck who was the first to give. Whatever was needed in the area of fundraising, Buck took the lead. And he involved his top people in Miller also.

Miller Electric and NECA

When it came to being "all in" with NECA, Buck said it early on: "Miller Electric is going to be part of NECA. We're going to support NECA and everybody throughout this organization will also support NECA."

I hear it echoed in the younger guys at Miller today, who say Buck gave them their marching orders. In their thinking, when it comes to the association, they're part of NECA and they'll use the services so they'll also support the NECA services. They said if they have an office in Dallas, for example, they'll get somebody there involved with the local chapter and also involve them on the board of directors if they can. The guys at Miller are not going to sit and just take … they'll participate in any way they can.

Strong Minds Discuss Ideas: Eric W. Livingston

I started at Miller Electric in 2008, working in the tool room, and later in the Fleet Department. Today I'm an assistant project manager in the healthcare division. I guess I can say I've known Buck Autrey, or Papa—as his grandchildren and great-grandchildren call him—my whole life since I'm his youngest grandson. My first memory of Grandfather was watching him replace the handrails on the deck at our old house at Lake Geneva. I was around two years old, yet I definitely remember the time he showed me how to use a hammer.

The fact that I can't come up with a couple of words to describe my grandfather says a lot about him. If I had to describe him to someone who didn't know him, I would say he was intimidating—which is what I usually hear from people who don't know him. Yes, he's a big man and he can be intimidating, but what they don't know is he's far more compassionate, considerate, and understanding than anything else. And my grandfather expects as much of himself as he does others. Considering his history, his position in the company, and the number of people he's employed, that says a lot about him. I think everyone who knows him would agree. I just can't remember a time he's wavered.

When I first thought about working for Miller, our nation was in the depths of the 2008 recession. Although my grandfather could have carved out a position for me at the company, he instead advised me to take the apprenticeship course. As I considered his suggestion, I clearly remember his stern advice. "If you're going to go through this," he said, "you've got to be the top in your class, or it won't be worth it."

With those intimidating words as my guide, I graduated top apprentice in 2013—the sole family member to do so since my grandfather, roughly sixty years prior. In recent years the company has been transitioning his responsibilities to the next generation, so I never had the opportunity to work with him on a project. I always imagined it would be both fun and educational. But since my relationship with him has always been grandfather to grandson, I guess the age and job position differences prevented any hand-in-hand interaction with him.

Grandfather is adamant about the separation of business and home life. If the subject of work comes up while he's at home, he'll silence it. He firmly believes when we're with our family we need to enjoy the time and leave work for the next day. I admit I'm not as successful with the rule as I could be; sometimes I bring my work home with me. I generally catch myself at it due to Grandfather's stance on the subject.

Given his love for the water and fishing, I'll use a familiar analogy to describe what he's done for this company. It's as if he took an old leaky boat, cleaned it up, shined it up, and turned it into a yacht ... better and more powerful than it's ever been. When he took over, this boat had a lot of unused potential, so he did what he had to and sacrificed to transform it into what he knew it could be. To be honest, that may have been harder than building a new boat from scratch. But he did just that, and put it back in the water and continued to trudge forward. He has continued to grow and shape our yacht to this day. No matter how it's stated, Miller Electric wouldn't be what it is today without his involvement.

I think anyone would tell you if my grandfather wasn't at work, he would probably be on the water. And if not on the water, he'd be traveling somewhere with my grandmother. It's tough to say which one he loves more—which brings me to my grandparents' relationship. I've always admired the strong commitment they have for one another and it's very rare you hear such things about a couple married for sixty-five years. In that vein, I'm proud my wife and I got together at a young age, as they did. I hope one day I'll be able to say the same thing ... that God willing my wife and I will have been together as many years.

Another way to describe my grandfather is through advice he gave me some time ago. Quoting Socrates, he said, "Strong minds discuss ideas. Average minds discuss events, and weak minds discuss people." My grandpa and I talked about that quote in depth multiple times, and I truly believe it defines him. He's never been one to sit around trashing someone's name. In fact, he doesn't even want to stay in the same room when someone offers up such talk. He believes there's a better way to utilize your time and your gray matter. As he said, "There's so much time in the day to do what we need to do, so why would anyone want to sit around and waste time talking negatively about somebody else?" He lives by this belief, and I try to carry out his example every day. I'll do my best to instill it in my boys too. Again, as Grandpa would say: "If you have something you want to say to a person, say it to their face if you must. Otherwise, there's no reason to talk badly about someone who isn't there to defend themselves."

It's hard to say if there was anything I would change in my grandfather, since he has so many admirable qualities. I love the fact that he's so open and compassionate with his great-grandchildren—our children—maybe more than he was with my generation or my parents' generation. My grandfather is a tough man who expects a lot out of you, but as I said, he also expects a lot out of himself. He practices what he preaches and leads by example. So if there is anything to change, I'd like there to be a time when he'd walk up and offer a hug or something. I'd hate to see him go out when it's all said and done with the same rough exterior because, even though we've figured out ways to get around it, I think we would benefit if we could embrace a little more. Nevertheless, with all the wonderful things about him, it's not much of a negative.

For some reason I still find myself striving to impress the man, like I did as a boy. My motivation and determination with which I approach work, fatherhood, and being a husband are shaped by the drive my grandfather instilled in me. I hope there's never a day when I'm not working hard trying to impress him. I truly believe he's driven me to accomplish a lot of things I might have never even challenged myself with. If there is someone in your life who makes you feel you have to do more and achieve more—though it may be stressful at times—you'll be further along than you would imagine.

Eric W. Livingston
Assistant Project Manager

EIGHT
Major Projects Drive Growth

Throughout Miller's rebounding years, our ability to land a few major projects helped shape our growing reputation, our profitability, and our net worth. Early on in the '60s and '70s, we were successful working with the paper industry and Ed Witt was mostly involved in such projects. For example, the Gilman Paper Company is one where he established a strong relationship with the manager. When that manager left he took Miller Electric—Ed—with him. Ed was so close to that manager that wherever they moved, Ed was there to oversee their electrical work. Gillman Paper was yet another instance of an ongoing project resulting from trust and a strong relationship.

Another major project was the Blue Cross Blue Shield building on Riverside Avenue, starting in 1972. We negotiated with the customer that we would be the sole contractor to bid the job, so they included us in their specifications requirements. Such deals became a highlight of my negotiating ability … convincing a customer to designate us as their exclusive electrical contractor. When we completed the Blue Cross project, we went on to secure their maintenance and expansion work, which continues to this day. They were one of our first major customers on the commercial end.

Other projects included the Jacksonville public school jobs. I landed many of those contracts before I became president, and we stayed with them over ten years, working on around thirty schools in all. In fact, estimating and wiring schools became one of my specialties. The jobs ranged from below $30,000 for a small elementary school, to over $400,000 for a high school, but we seldom did a job over $300,000. Nevertheless these projects were many and helped our cash flow at the time, although we seldom cleared more than 10 percent on any job.

When we acquired the Paxson Electric Company we took over some big jobs, like the University Hospital worth several million dollars. We also assumed the Baptist Hospital account and have worked on every Baptist Hospital job in Jacksonville ever since. We continue to do their hospital and facility maintenance, including new expansions for Baptist South, Baptist Beaches, Baptist Hospital at Yulee, and the new Baptist facility at the downtown children's hospital. In fact, today we're working in every major hospital in Jacksonville … quite a deal when this city is known for its healthcare industry.

One of our long-term projects was the Savannah River hydrogen bomb facility in Augusta, Georgia. Miller Electric held the maintenance contract since the early 1950s, and every year we had to renegotiate. We were limited to how much wiggle room we had, in accordance with the wishes of the owners who told us they would pay us so much and that was that. I was the one who handled the negotiations every year until 1990. Eventually they moved all of the contracted labor onto our company's payroll, which gave us additional labor fees on top of the rest. It was a good arrangement, until suddenly the DuPont Company—the general contractor—pulled out. Within a year the replacement company got rid of us, and the Bechtel Corporation took over the payroll. That was the end of a very satisfactory forty-year run.

I can actually say losing the Savannah River contract was a good thing. It freed me up to work full-time with our company and pursue long-range opportunities. Until then I had to go to Augusta several times a year—for over twenty-five years—to take care of the trouble spots and negotiate with the unions. The silver lining of losing that contract came not long after, when we negotiated the takeover of the Paxson Electric Company … which virtually doubled our business overnight. That new business meant we never had to worry about going to Augusta again.

Some of our recent jobs were data processing centers like the Bank of America in Richmond, Virginia. I believe that project was one of the largest we've ever done and it amounted to $50 or $60 million. Data centers are now among the larger jobs we do and they include uninterruptible power sources, massive generators, and extensive prefabricated systems. We also produced the Blue Cross Blue Shield data center at Cecil Field in Jacksonville, one of our first large data center projects.

Awarding prizes at the Savannah River plant

A Word about a Sense of Humor

You can't be a good contractor without a sense of humor. Believe me, there are so many characters to deal with in this business that without a sense of humor we'd go nuts! Not only that, if you want to be a good salesman—which every project manager should be—you must also have a good personality. And you can't have a good personality without a sense of humor to go with it. That's part of being a complete manager ... being able to laugh when somebody pulls a real dumbo. It's all right to make a joke about something that happened, within reason. A manager knows that anyone who has been around for a while has done something just as stupid at one time or another, so making light of a mistake tells a person it's not a bad thing. Instead, say something like, "As a matter of fact I did the same thing ten years ago!" That will go a long way to help that person think more before he does it again.

All In: Henry Brown

I always knew I wanted to be part of Miller Electric Company. After all, growing up as Buck Autrey's oldest grandson, how could I ever want anything else?

I started in the summers during high school and college, working in different positions: the tool room, the accounting department, the estimating department, and even some time in the field. After graduating from Stetson University I began my professional career as a CPA with Deloitte & Touche, and later became a financial analyst with Atlantic Marine. The time outside of the company provided a valuable learning experience, but it confirmed even more that working with my family at Miller Electric was where I belonged.

In April 2001 I joined the company and worked for my mother as the assistant controller. As the company grew and I learned more about the industry and the operations, I took on additional responsibilities. I moved from controller to vice president, and later senior vice president of corporate

---

affairs. In September 2012 I was given the opportunity of following in my grandfather's footsteps as CEO of Miller Electric Company.

I've always been close with my grandparents. Growing up, we lived around the corner from them and always spent a lot of time at their house. I feel very fortunate that we had such a close relationship and have learned so much from my grandfather over the years.

My grandfather—I call him Papa—is one of these larger-than-life individuals. He's always been outgoing, gregarious, and the kind of person everyone wanted to be around. From a work standpoint I would describe him as hardworking when he needed to work, and hard at play when he wanted to play. That's something I admire about him: he always had good balance in his life between work, family, and play. I'm sure he was not as much that way while he built the business, but he has definitely made room in his life for family and fun since then.

Papa has always been a huge influence in my life and he has always been my greatest mentor. He is someone I always asked regarding what I should do in different situations, such as where to go to college and what degree to pursue. I remember what he said about majors. "Whatever you major in," he said, "after you graduate be aware of what kind of job you're out of. Know that you're an *unemployed* accountant, or an *unemployed* lawyer, or an *unemployed* whatever it might be."

That was good advice that I think most young people don't take into consideration. I majored in accounting and although I spent a short time in that field, it gave me a professional degree and a certain amount of credibility in the business world that would not accompany a young professional with most other degrees. I think that advice also gave me a significant head start on my career and contributed to my early success. That's what Papa meant about having credibility after you graduate … something that people, when they look at you, will say, "I can call that person to do something and know he has the skill set to do it." I think such wisdom came from his background as an electrician … whatever happened in his life, he always knew he would be an electrician. He had the skill set, the background, and the credibility.

Before I joined the company, Papa really wanted me to be sure it was the right thing for me. I won't say he discouraged me from coming to work at the company, but he certainly wanted to make sure I had the opportunity to do other things first. I probably would have worked longer outside of the company, but the right opportunity came at the right time for me to make the transition. That was another piece of good advice I should apply to the next generation: Start your career outside of the family business to build your credibility and experience. Then when the time is right, join the company for the best reasons.

When I was little I remember Papa's down time was mostly spent at their lake house, either in the garden or out fishing on the lake. He also played a lot of golf. I remember going to the driving range with him a few times, but unfortunately never had the opportunity to play a round of golf with him. I don't think he had the patience to play golf with kids, anyway, and he gave up the game before I was old enough to join him on the course. I remember how methodical he was at the driving range: his swing, the setup … his whole approach. You knew he was good at it! Practicing at the range today I can still hear his voice helping me with my grip, my stance, and ensuring I knew my target. "Know your target. Don't just swing away!" … Another piece of advice for both golf and business.

Around the early 1990s, Papa made the transition from golf to sport fishing. His peak fishing years came when I was in high school and later Stetson, so I didn't spend as much time fishing with them as my brother Daniel, who was four years behind me. He was at the right age to be fishing with my grandfather and they developed a strong bond during their time on the water. It is amazing how much those early years with him influenced both Daniel and me. Today, Daniel is still an avid fisherman and I am still an avid golfer, although I don't think either one of us will ever match Papa's level of talent or passion for either sport.

A quality about Papa I think is great is whatever he did, he was all in. He has a passion for whatever he's doing … whether building his career, growing Miller Electric, or his involvement with NECA. He was never *just* going to be involved with NECA … he was going to lead it. Of

Susan Walden with sons Henry Brown (left) and Daniel Brown (right)

course in short order he became president of the association. It was the same with golf. He wasn't simply going to play the game, he was going to be a championship golfer. He dedicated himself to that and won several amateur tournaments. After he gave up golf he started sport-fishing, and he wasn't going to do it just for fun; he was going to be the best sport fisherman around. He won many tournaments and multiple boat-of-the-year awards. Whatever he did, he was fully committed.

As passionate as he was for winning, he never lost sight of the bigger picture, his family. He always put them first and his generosity is often limitless. He also knew his greatest responsibility in the company was to leave it better than he had found it. When Mrs. Wynn promoted him to president of Miller Electric Company, she trusted him to be the steward of the company and her father's legacy. Papa knew it was his responsibility to grow and protect the company for future generations. That responsibility of stewardship was the root of his long-term thinking that has made the company and our family what they are today. It is a responsibility he has passed on to the next generations, and it is a responsibility that we are proud to carry on.

Papa has been a huge influence on my life. I am so thankful I have been able to spend the last fifteen years working with him, and thankful for the trust and confidence he placed in me to lead the company into the next generation. I wouldn't be the man I am today without his advice, his generosity, his leadership, and his love. Like so many others, I owe everything to him.

Henry Brown
CEO

NINE
Guides, Encouragers, and Mentors

I mentioned a couple of times before that when I married Betty, my life completely changed for the better. I have to say much of the change also began with her father, Roy White. Since my father-in-law had been in the trade a long time, his presence in the company helped me in a couple of ways: first, to get my foot in the door, and later, to advance in the electrical industry. Roy had an excellent reputation with both the company and the union, so his name alone opened many doors for me. Of course, as I grew into the electrical contracting business there were several others who, through their education, technical expertise, and good character, helped me along the way. Here are some of them.

Bill Binkley

In the early 1950s at the start of my career, our local NECA chapter manager, Bill Binkley, became a real friend and mentor. Bill was also a friend of my father-in-law, Roy, since their families grew up together on the Northside of Jacksonville. After I graduated as Outstanding Apprentice, Bill started to follow my career. Later, when I was a journeyman he encouraged me to become part of the teaching staff for our apprenticeship programs, which I did. He continued to follow my career in the field and also after I came into the office.

When I'd been teaching for some time, the administration asked me to give a speech at the 1962 apprenticeship graduation ceremony. Bill, who was still the chapter manager, introduced me to the class as their speaker. It was a good talk except for his introduction, when he called me the "vice president of Miller Electric Company," which was a mistake, of course. After the laughter died down, I gave my talk and it went very well, I must say.

His blunder came back to bite me when someone told me they overheard Mrs. Dandelake, the president's wife, make a comment to a friend in the ladies' room after my speech.

"He's not a vice president," she said, "and never will be a vice president. My son is going to be vice president!"

After hearing about her comment, Bill reportedly told Jim Dandelake what he thought. "Well, if Buck's not a vice president, he *should* be."

Bill constantly urged and pushed me to do things I wouldn't ordinarily have done, such as teaching apprenticeship. That experience taught me a lot about myself and I believe I became a sharper person as a result. I also learned how to focus better, which came in handy after a busy day when I had only an hour to prepare my lesson plan. I enjoyed teaching the electrical trade to the young apprentices, and attribute those discoveries and personal growth to Bill Binkley's encouragement.

John Geubeaud

Another man who influenced my life was a local union business manager (there were some good ones out there) named John Geubeaud. John was also a friend of Roy White and John also turned out to be a good friend of mine. He aided me greatly in my transfer to the inside union after my friend Slim was killed. And like Bill Binkley, John followed me through my apprenticeship years.

Andy Bernard

Andy took over as NECA chapter manager in the early '60s, after Bill Binkley retired. Andy worked for the Florida Department of Labor and headed the state's apprenticeship training programs. He observed my work in the apprenticeship training and gave me a lot of assistance during that time, and beyond. He was a good follow-on advisor after Jim Dandelake became ill and I became president. He was also there for me when I was elected president of the local NECA chapter, and was among the first to encourage me to run for vice president of NECA.

After I became a governor for NECA, Andy and I traveled a lot together to various governors' meetings with the managers. He encouraged me to run for vice president, and also helped gather votes to get me elected. After I was elected, he hosted parties and events to promote me as president of NECA, lining up several governors to support my bid.

In addition to being my gifted campaign manager, Andy was a good supporter, a personal friend, and a friend of my family. His wife and Betty also got along well, and both of our families traveled together quite a bit. Unfortunately they're gone now, but through those fleeting years we enjoyed a great relationship.

Bob Colgan

Bob was the national president of NECA, and I'll never forget when he was elected. It was 1973, when I was NECA's incoming vice president and several NECA Southeastern District governors met at Las Vegas to decide how we'd vote for president. Two people were up for the job: Chuck Scharf from the L.K. Comstock company in New York—one of the largest contractors in the country—and Bob Colgan, a medium-sized contractor out of Toledo, Ohio. The meeting was lively and the outgoing vice president from District Three tried to convince the governors to support Chuck Scharf. When he sat down, Charlie Fulton from Atlanta stepped up to the mike.

"Quite frankly," he said, "you're the outgoing vice president, but we have our incoming vice president, Buck Autrey, right here. So I'd like to know how he feels about who he would work with better."

I stood up and gave them a spiel I hoped would sway them to favorably consider Bob. "I do believe Bob Colgan would make an excellent NECA president, and I'm sure you all know that he and I go way back. To answer Charlie's question, I have to say, like Bob, I too am a small to medium contractor, and I'd feel somewhat intimidated in a relationship with a company the size of Comstock. With all due respect for Chuck Scharf, I'd feel better and more comfortable dealing with another company of a similar size as mine. Further, I believe Bob Colgan has the right message for NECA and what the association needs today, so I'd appreciate it if you would support Bob."

Apparently my speech hit home. Bob won by the exact number of votes represented in our meeting that day, and he knew it. Bob was another one who persuaded me to run for the NECA presidency.

Bob Higgins

I count Bob as one of my best friends for life. Having said that, we were still at loggerheads most of the time due to our differing management philosophies. In his favor, though, Bob was the paid director of NECA and he had to manage the people the best he could.

We traveled a lot together and every time we boarded a plane I'd pull out a deck of cards to kill time playing gin rummy. He was a slick card player and I enjoyed it too, even if I wasn't so good at it. I kept a record of the wins and losses and after six years of traveling and playing gin rummy together, I finally tallied the scores. The result was I owed him $3.50. No, we didn't play for high stakes … probably about a half cent per point.

One year Betty and I drove our motor home to Toledo, Ohio, to visit Bob Colgan and his wife, Emily. On the way, we stopped in Washington State to see Bob and Carol Higgins and spent time with them in their home. Later we piled into the motor home, drove to Toledo, picked up the Colgans, and wound up on an island in Lake Erie, where the six of us enjoyed a long weekend.

Bob owned a cabin on a tributary to the Choptank River on the Eastern Shore of Maryland. We enjoyed many weekends during duck hunting season over several years, and I even have a few trophies of several ducks I killed, which Bob had stuffed for me.

We also traveled overseas to places like London and Edinburgh, Scotland. One time we rented a car and drove to St. Andrews for some golf, and of course driving in the UK means everything is the opposite. Bob was the driver, so he sat on the right side of the car, driving in the left-hand lane, while I sat on the left side of the car where the steering wheel *should* be. Yes, I felt completely out of control. Those Scottish hills with sharp curves and no guardrails separating us from a long way down a cliff kept me white-knuckled for the entire trip!

With Emma and Abbey - 2008

Buck & Betty, Daniel & Jacqueline, Susan & Ray, Henry & Jennifer, 2010

NECA's Academy of Electrical Contracting annual event at the Broadmoor, Colorado Springs - October, 2010

A pig roast after a long day of bird hunting in Argentina

Quail hunting with Howard Shaw

Fly-fishing in Goodnews Bay, Alaska

Dove shooting in Argentina

In Alaska with (L to R) Buzz, Matt Kenyon, Daniel, Dan MacArther, and Ron

Left to right: With Henry Beckwith, Rick Cox, Vaughn Beasley, Buz, and Ron - 2012

Bob was a master travel agent along with everything else. One time during my presidency, Betty accompanied me on an overseas business trip. We flew to Nice, France, and drove down to Monaco for a meeting. Later we traveled up to Bern, Switzerland, and after a break, headed over to Paris. I don't know how Bob got away with paying for the next leg, but he arranged for us to fly from Paris to New York on a Concorde jet.

"Let's do it," he said. "We probably won't have another chance again—and let's hope I don't get fired in the process."

If I remember correctly, the only seat they offered on the Concorde was first class, and every seat costs around $12,000 for the three-hour trip between Paris and New York's Kennedy Airport. Somehow Bob managed it and I don't think anyone ever questioned it.

Jim Dandelake

I would be remiss if I didn't mention Jim Dandelake, Miller's former president. Jim was an intelligent man and excellent engineer who passed a wealth of estimating knowledge on to me. He taught me a lot more about our trade, including the basics of calculating labor units. His system eventually influenced our national association's curriculum.

The evolution of his method started in 1956, when I was a superintendent at a large paper mill job. We introduced a time and motion study where we developed unit prices for the time it took to do a job from one point to another. We defined each job segment and turned it in on time cards. Someone in the office correlated the data into codes, which meant we kept time according to a coded system. That way we knew how much time it took, for example, to run one hundred feet of conduit of a certain size. We then developed our own list of labor units, which NECA later adopted into what they call "The Manual of Labor Units" by the National Electrical Contractors Association. Jim Dandelake was a major contributor to NECA in that way.

Jim also encouraged me to take those courses in engineering through the International Correspondence School since he was a student himself. As mentioned, I took his advice and completed two divisions of engineering through ICS.

If Jim influenced my electrical education, he also influenced in a somewhat negative way as to how I would run Miller in the future. I say that because Jim made an awful lot of mistakes in the way he ran our company. He was a brilliant engineer who, unfortunately, made bad business mistakes and poor personnel decisions. He also made poor decisions about the kinds of jobs he took on, the *Caribe Queen* being one of many.

Not to end my recollection of him on a negative note, Jim stayed with the company a long time and contributed much to its existence. I learned a lot from him and will always appreciate his influence in my profession.

A Word about Employee Recognition

We honor our top producers each year for their outstanding performance in safety, production, profits, clerical work, or any other part of the business. Not only do they get recognition in front of their peers, along with a plaque ... they receive real profits. Personal recognition is important, although management doesn't go around constantly slapping them on the back. Their good work is made apparent through close communication by supervisors, customers, and peers, so at the end of the year they will know they've done a good job. Now that I'm in my semiretirement years, however, things are different for me. Whenever I'm in the office, all I do is give praise. I don't criticize anymore, but instead tell them when they're doing a good job. After so many years on the job, it's my privilege to lay it on thick.

Uncle Buck: Ed Witt Jr.

My dad, Ed Witt Sr., worked with Buck Autrey for over fifty-eight years, and as far as I can remember, I always called him "Uncle Buck." That is, until I started getting my paychecks from Miller Electric. Then he became "Mr. Autrey" around the office. One of my first memories of Uncle Buck was at Lake Johnson in Keystone Heights, where my family owned a vacation home next door to the Autrey family. I spent my early summers at the lake, where I loved to go fishing with the two of them. Uncle Buck and my dad had a real passion for fishing, and they went as often as possible, every morning and evening. I tried to go with them whenever they would let me. The lake water was high and we caught almost everything we wanted.

One day when I was five or six, we returned from a fishing trip and I stayed on the beach while Buck and my dad walked up to the porch. Buck called down to me and said if I brought them a beer from the boat's cooler I could pick any lure I wanted out of his tackle box. I came up to the house with the beers and the lure of my choice, and Buck made a big deal about how I picked the *best* lure out of the whole box. True or not, he made me feel as if I'd done something extraordinary.

Buck's unique way of encouraging people has been the same throughout my career. He has a way of giving you a sense of "job well done" without having to say it. He was bigger than life to me, not only in size and his booming voice, but with the degree of respect I had for him. Even at my young age I knew he was somebody very important to my family.

In my late teens I became a golfer. Buck was a tremendous golfer at the time, and I'm sure he somehow influenced me to take it up. Although we didn't play together often, when Dad and Buck participated in The Players Championship Pro-Am event (in the early days), I caddied for them. Like everything else, Buck was intense with his golf game, yet he always made me feel I was part of the team.

When Buck did something, he did it all the way. And when he was done, he was done ... kind of like Forrest Gump: "Okay, I'm through running now ...." And he didn't run anymore. In the same way, when Buck started something—anything he chose—he was very good at it. After a while when he felt it was time to quit, he stopped. It wasn't as if he drifted away or his golf game went downhill. No. He stopped at the *top* of his game. Whether playing golf, game fishing, NECA, or Miller Electric, he's definitely all in or all out.

After I came to work at Miller I had the opportunity to join Dad, Buck, and a group of their peers and business associates on hunting trips abroad. These trips were part social and part business. We hunted in Argentina, Uruguay, Mexico, and other South American destinations. At that time in my life I was made to feel like one of Buck's friends or peers—even though I knew I wasn't. I felt more engaged in the group and got to know a lot of stories … stories we now share together.

One night in Nicaragua during one of those hunting trips, Buck's associates pulled one on him with the help of the lodge staff members. We were at dinner and Buck sat at the head of the table, as usual telling stories about growing up. The dinner was an entire roasted pig, and prior to serving dinner a story circulated around the table that it was a great honor for someone to receive the head of a pig at such an occasion. So the staff said they'd dress the pig up just for Buck. They laid the pig on a platter, garnished it, stuck a big ol' cigar in its mouth, and covered it with a silver lid. When they set the platter in front of Buck and pulled the cover away, there was the head of the pig … all roasted and beady-eyed, with that big Churchill in its mouth. Of course, Buck erupted into a belly laugh and took it in stride. I called him Uncle Buck on those trips, but I knew to draw the line in the office, where I called him "Mr. Autrey." Now that he's semiretired, the "Uncle" occasionally slips out.

I'm fortunate to have seen Buck, or Mr. Autrey, wearing many hats. I knew him as a friend of my parents, as one of his virtual family members, the ultimate boss at work, and in some ways as a personal friend. I felt I knew him better than other employees and possibly better than even his friends, since I saw him in every other aspect of his life. The greatest thing about seeing him from those perspectives is I know deep down he's one of the most caring people I've ever known. At work his outward persona—according to my coworkers, anyway—can be intimidating. Frankly, I was also intimidated by him because of his position; yet my interaction with him at work was never all that threatening. It helped that I was always honest with him because honesty and my deep respect for him allowed me to be vulnerable without fear of repercussions.

Buck used a quote I've often repeated, and rumor has it he always said it when we didn't make a profit or weren't successful in some way. "It's cheaper to do nothing for nothing than something for nothing." In other words, he'd rather you not do anything than do what you did which cost the company money and wasted a bunch of effort to do it. Still, Buck has fostered a culture in Miller Electric where people have room to fail without fearing negative consequences … as long as they're aware of their mistake and genuinely try to fix it.

When it comes to open communication, Buck always wanted to know what was going on. As long as he heard the truth there wouldn't be a problem. Failing is one thing, but not knowing why or not caring enough to prevent it from happening again is what gets people in trouble. At Miller we've always had the freedom to fail, unless it becomes a habit. If it does, it becomes obvious the person didn't really learn anything. Such values have been developed over many years, and as one of the leaders of the company I'll help keep them alive with a modern, up-to-date spin.

Occasionally I saw Buck slip out of character. One such time happened ten years ago when the company held an executive retreat and some of us had to make presentations to the other executives. Mom was very ill in a Jacksonville hospital at the time, and the retreat took place down in Palm Coast. I attended, but my dad could not because he was having some issues himself—taking both of my parents out of commission.

During my PowerPoint presentation I stumbled once or twice, and then started to cry. I stopped what I was doing and spoke to the room: "You know, I really shouldn't be here." I sat down to gather my stuff and looked over at Buck. He was also crying. He said something to the effect of: "Ed, go ahead and be with your mom. You need to get out of here. … I'm so sorry this is happening."

Buck's kind words and action validated the relationship between our families. The Buck I knew at work was more of a hardworking authoritarian, and to see that side of him touched me deeply. I gathered my stuff and left. A month later, Mom passed away.

Another time was when Mrs. Wynn died, in her late nineties. I spent a lot of time with Mrs. Wynn when I was younger, while Buck went on NECA trips as their president. At the time my dad was an official of our local NECA chapter, and sometimes Mrs. Wynn traveled with us to various meetings or events. We often tagged along to escort or help her and do whatever she needed. I also believe she had a huge impact on Buck's life. Years before, she was the one who said, "Buck, we can close the doors or you can come in here to help me get it going again."

Mrs. Wynn had no close relatives, so Miller became her family. Her husband had passed away and she didn't have any children—only a few distant relatives. Buck gave the eulogy at her funeral. It was the most heartfelt, emotional talk I've ever heard. I believe he even referred to Mrs. Wynn as "one of the loves of his life" while he wept. Again his words and actions validated what I knew of their relationship.

The future of Miller Electric will be affected by many things. I can't say with any certainty what's going to happen with the economy, but I'm confident that by maintaining our core values that formed the basis of our company when Buck took over as president, we will ensure our foundation won't be lost.

Ed Witt Jr.
Senior Vice President of Construction

TEN
Moving Ahead

Succession

Miller had no long-term plan of succession, other than Mrs. Wynn knowing the company had to continue in her father's name. At the time I took over the company we were in survival mode, so there wasn't time to think about the future. All we tried to do was make the future better each year. Aside from that, the only tangible carryover was the company shares Mrs. Wynn bought from Mr. Dandelake when he retired and passed them on to me. Once we became solvent to where we paid our bills on time, I began to look further ahead. Then in 1972 I started a profit-sharing program and it got better each year.

In the early '90s after we took over the Paxson Company, Mrs. Wynn and I agreed I could give the children some of my stock. I suppose that's when I started thinking about succession … how I could take advantage of the present tax laws to give my children the stock. By then I already knew who would be the successor because of the way the family worked in the company. I knew it would be Susan and Ronnie … they both were dedicated and hardworking. Since I had already figured out who the successors would be, it was a matter of working out the stock distribution.

Since I was a minority shareholder, I was able to valuate my stock at a reasonable price and transfer it as part of a lifetime gift tax exemption. The reduced value brought it in below the amount allowed for a lifetime exemption for gift and an estate tax. So I gave a lot of our stock to two of our children, Ron and Susan. I also gave Ed Witt additional shares, until the three of them owned equal portions. I held on to my larger portion of stock.

Buck's 80th birthday - 2012

Grandparent's Day

At a recent Academy meeting in Colorado Springs

With the four children and their spouses - 2012

Family Photo - 2012

With Betty and our "four doorsteps"

Papa with Walker at the helm

It stayed that way until Mrs. Wynn couldn't work any longer. Prior to her death we bought her shares, which made me the majority owner of the company. After they changed the law to allow a greater amount of stock to be given away, I shed more of my shares to Susan and Ron. Later on, the two bought out Ed Witt's stock upon his retirement. Sometime later they bought out my stock.

When Ron became CEO, he and Susan became 50-50 partners. They were the only two stockholders aside from the 1 percent ownership I retained … to keep the arguments down. I became the leveler, so to speak, the arbitrator of any discussions that came up. And there were a lot of them, which is expected when different members of a family run the business.

Sometimes it's not easy to bring children into the business since they have their own families to think about. They want their own children to be involved, but sometimes the children's skill sets don't match the needs of the organization, which makes it difficult at times. After Susan and Ronnie became full owners of the company, Susan transferred her stock into a family trust, and later on the trust bought out Ronnie's stock when he retired. At the time, capital gains tax was 15 percent, and the next year it was raised to 20 percent, plus Medicare payments. In all, there would have been a 23 percent tax on it if they had waited. That's how my long-range plan came about for our children.

I have to mention another part of our financial succession, which is the stock options and profit sharing. It has always been my belief—and I mentioned this already—that employees should be given the opportunity to share some ownership in the company. We established that system some time ago, and as of today it has worked out quite well for everyone.

A Word about Inspiration

I think inspiration comes mostly from observing other people, particularly in the workplace. If you see someone doing a good job and you observe how they're doing it, you might be inspired to do it the same way. Or if a person is doing better than you in a particular area, find out how and go on to improve yourself. The other way around is if you want to inspire people to do things better, do things better yourself. Which is how I became president of NECA … I used my position at Miller and our track record to inspire others to support me in a leadership position. Peer pressure can also be inspiring. Let's say you want to raise money for charity and you ask yourself, *What's the best way to do it?* The answer might be to convince someone people respect to give money for the cause, and show that person as an example for others. If they admire that person to any extent, others should be inspired to be in the same pot with him, so to speak.

The Best Kind of Fear: David Long

In 1975 I found a job as an entry-level worker at Miller Electric Company. My dad had worked in the field with Mr. Autrey and Mr. Witt, and told me if I wanted to work in the electrical industry I should join Miller. In those days I didn't have any idea who Mr. Autrey was.

As I began to grow in my career as an apprentice, I found out Mr. Autrey was held in high esteem as a man of vision. I also heard he completed both the inside and outside apprenticeship programs almost simultaneously—nearly unheard of in those days. After completing his apprenticeship, he went on to teach apprenticeship himself. Early on I learned Mr. Autrey believed every individual was responsible for their own career, for making themselves and their family better, and for bettering their company.

After I completed the electric apprenticeship program in 1981, I was given the opportunity to run a service truck at Miller. In those times I would see Mr. Autrey occasionally in a hallway. And yes, there was a level of fear because I didn't know him personally, but I knew what he stood for. In 1992 when I came from the field into management, I had an opportunity to sit down with him in his office. That was my first conversation with him and it was frightening.

He was methodical in his approach to my opportunity, and acknowledged what I had accomplished in the field. He was also complimentary of my customer skills and the money I had made for the company. He was also very challenging, saying this was an opportunity he had embraced once in his own career, as did many others in the company. He said my new job was not an impossible task and assured me I was not being set up to fail, but to succeed. He reminded me there was no time clock at the front of the office ... which I thought was unusual. In the field I always had to turn in my time and was responsible to account for every minute of every day.

Noting my look of confusion, he said again: "Son, there is no time clock at our front door. You're going to be working in management and you will be your own judge, so rest assured I will know how hard you work and I'll know how many hours you put in by the numbers you produce and the results I see. So this is your time clock, son. A lot of people can't manage that way, so you will have to understand self-discipline."

I started out estimating $4,000 and $5,000 jobs, and as my career grew I had more interfaces with Mr. Autrey. I remember the first time I was invited to his home for a Christmas event with him and Miss Betty. I've never felt more welcome or more a part of the organization!

Next I became a project manager and then a senior project manager, learning more about responsibility and accountability. Mr. Autrey was not a micromanager, so he was not in my office every day wanting to know what was going on. He simply wanted to know the results and the answers. He expected the truth, accurately, every time we spoke. As Mr. Autrey used to say: "Son, you are not always going to do well. You're going to make mistakes and you have to learn from them. And in order to learn from them you must face the reality of understanding what the mistake was."

Sometimes my mistakes were self-induced and sometimes they were out of my hands. In every case, I learned.

I can say Mr. Autrey's number one quality is complete honesty. Number two is his strong character, and number three is his ability to think about the good of all. He always felt if we did good, he did good ... it was never about him. I remember as a young manager, when the company grew geographically and I had to travel a lot. Since I was gone so much, my wife and I decided she should be a stay-at-home mom to take care of our son. I told Mr. Autrey I worried about the arrangement because financially it took one of our incomes out of my household. Mr. Autrey said, "Son, it has a way of working itself out." Sure enough, when the next year rolled around, my pay was increased enough to offset what we lost. He believed if I was out there giving everything I had, he would give a portion of what he had to me and my family.

In time, Mr. Autrey, Mr. Witt, and Ron Autrey gave me an opportunity to become a vice president of the company. I remember how they mentored me on my new job while on trips or during lunch. Mr. Autrey also advised me on how to act and how to carry myself. It was necessary, and after I became an officer in the company and people saw me, they saw Miller Electric, not David Long.

Mr. Autrey believes education is the only way the company will move forward. He feels that education produces individuals who believe in themselves and are confident with others in the company. He said no matter what we do we must give back and make others better, and in the process we naturally will become the best. He once said we could judge the effectiveness of our education program by how people in the field felt about themselves, how they performed, how our customers viewed our work, and how our competitors viewed our products. I took his advice to heart through the years.

Today I serve as District Three national vice president of NECA, co-chair of the National Electrical Training Alliance, and chair the NECA Workforce Development Committee. Every meeting I attend and every association event—no matter how large or small—I'm aware of the influence Mr. Autrey created years ago. His charisma, integrity, and respect for others led Miller Electric to greater heights and also affected the entire industry. Throughout his career his leadership benefited our company, the association, the union, and other contractors across America. Everywhere I go, whether California, Texas, or Ohio, people refer to him. The older generation ask, "How's Buck?" while the younger ones ask, "What would Mr. Autrey say about this?"

Sometimes I'm pulled aside and told a story about Mr. Autrey ... about "How Buck did such and such," how he affected the association, and how he created new paths for people to follow. I can tell they truly consider him as a partner, not only a peer. They could pick up the phone and call Buck Autrey any day or night. They knew he would give them honest, truthful answers about any situation. They understood his ultimate goal was to make *their* company as good as Miller Electric. I've had many conversations with John Grau at NECA, whom Mr. Autrey interviewed for the position of CEO. John told me when he started, Mr. Autrey mentored him and he knew he could call on him anytime to get the answers he needed.

John also said Mr. Autrey worked with him on political situations in IBEW, the organization in which I later served as a National Labor Relations Task Force member. Many older gentlemen of the IBEW, and many others including their international president, often told me, "You make sure you tell Buck and Ed I said hello."

There is an unbelievable degree of respect there, even if they were on opposite sides of the aisle. And there was always a consensus in IBEW that Mr. Autrey wanted to work with more people to produce more man-hours and have a better wage for the worker. They also knew he wanted to create a healthcare package to take care of their families and a retirement fund they could live on until their life was over. That's what he did.

Mr. Autrey's activism in the electrical industry went far beyond Miller Electric. He created a ripple effect in the industry which ultimately influenced electrical contracting throughout the country. The older generation remembers how he went to war, so to speak, on important issues that changed the industry as we know it. Today's "middle generation" of the industry seeks leadership, answers, and understanding. So they come to people like me, Henry, and others, asking, "How do you do such and such?" Well, they know we know how to do it because Mr. Autrey, Mr. Witt, and others in leadership had the wisdom and the tenacity to get things done and teach us. The new, younger generation listens to those industry stories passed around by the 4,000 electrical contractors and

250,000-plus electrical workers. As I travel the country, there is no doubt Mr. Autrey is known in every city and in every state that has anything to do with the unionized sector of the electrical industry.

Mr. Autrey said in order for us to survive as a company, we must understand the union issues and be involved in them. He advised me to keep my friendships with both industry management and union leadership. The reason was they had families to feed, workers to protect, and they too had a job to do. Historically, Mr. Autrey was confident with the situation and often said: "If you're going to be responsible for labor relations, you should have as many people in the union on your speed dial as you have in your management organization. You need to understand labor relations, and that you're not always going to win. You have to understand they also have to win sometimes." That kind of leadership is what allows me today to walk into the IBEW offices with a high level of respect. Miller Electric is viewed as a company concerned about the betterment of both organizations.

David Long
President, Miller Electric

New Year, 2015, with Ray and Susan

Epilogue
Moving Forward on a Foundation of Trust

A Word from Henry Brown, CEO

When I read through the preceding chapters about my grandfather, I was reminded again of the tremendous legacy he passed on to his family and Miller Electric Company. His professional and personal life exemplifies integrity, innovation, and trust during a time of significant changes in the electrical contracting industry. As a result of his legacy, Miller Electric Company stands as one of the bright spots on America's professional landscape to this day.

I would say, of all his traits, his ability to build trust shines above everything else. After all, trust is the foundation of true successful business relationships, and Miller Electric Company is no exception. An early example of his foundation of trust is exemplified by Mrs. Jane Wynn, the daughter of our company's founder, Henry Miller. Fifty years ago, Mrs. Wynn's decision to place the company in the capable hands of a young executive named Buck Autrey reflected a high level of trust that protected her father's legacy and affected the lives of employees and their families for generations to come.

Even in later years when my grandfather had full control of Miller Electric, he was aware the company was much bigger than he was. He knew his role: lead responsibly, grow the enterprise, provide jobs for employees, reinvest in the company, and give to the community. His role was not to take everything Miller Electric could give him. Instead, he gave the company everything he had to give. My executive team and I are doing our best to continue this legacy of trust. We know we have been entrusted to build on the past eighty-seven years of establishing trusting relationships with our employees, clients, and suppliers. As a result, we have gone further than any strategy or well-thought-out business plan could ever take us.

In a 2013 survey conducted by Merrill Lynch, people over forty-five years old were asked: "What is the most important thing to pass on to the next generation?" A full 74 percent of respondents answered: "Values and life lessons." The response, "Financial assets or real estate," came in last. Another example came from Chris Heilmann, chief fiduciary executive of U.S. Trust, who said, "I've been in this business for forty-one years and from my experience, if wealthy people are faced with a choice of being able to hand down their money or their values—not both— they'd want to hand down their values." Put another way, Warren Buffett, CEO of Berkshire Hathaway, said, "… in looking for people to hire, you look for three qualities: integrity, intelligence, and energy. And if you don't have the first, the other two will kill you. …" Trust and integrity go hand in hand.

Although Miller Electric has always been a family business, it was not only the Miller family but the Wynn family, the Autrey family, and the Brown families as well. Our company today includes several families who have so much trust in the company that they encourage other family members to become a part of our team. Together we are creating the Miller Electric Company family that will last beyond our lifetimes. It all came through trust, the glue that keeps us together. Through this legacy our team knows our role is that of a steward, to maintain the legacy of previous generations by leading the company to a better place than it was when we found it.

Trust Powers Our Future

When trust is a key component of our relationships, what follows is the freedom to innovate and move ahead of the competition. Innovation powered by trust is shaping our future at Miller Electric Company. We bring a wide range of innovative solutions for our clients, including intelligent lighting technology, security solutions, intelligent electrical infrastructure, and even building models integrated into intelligent hard hats. The trusting relationships we have with our manufacturing and supply chain partners allow us to capitalize on the digital revolution and bring integrated solutions to our clients. Through trust, we can bring together solutions that are not stand-alone but, rather, built on each other, while capitalizing on converging technologies.

These are exciting times for our industry—times when energy, information systems, and facilities converge to create a new world of possibilities. Buildings, hospitals, factories, and cities are becoming more intelligent and more interconnected every day. We are approaching a time when everything will be connected to network-producing data to drive performance. When physical assets are equipped with digital sensors to capture, process, communicate data, and even collaborate with each other, they create game-changing opportunities in the form of reduced downtime, increased productivity, and energy optimization. Energy optimization can come in many forms, including effective use of renewables, energy storage, grid optimization, and of course improved energy efficiency.

Innovation in the energy sector is not only based on solar, wind, gas, or coal, but also in the convergence of the digital and physical worlds. Whether it is referred to as the Internet of Things, the Internet of Everything, or the Industrial Internet, data produced from the increased connectedness of devices and the insights developed from such data have the power to transform our energy economy.

At Miller Electric, we are excited about the possibilities of a more connected future. While we have helped customers with their energy infrastructure for generations, the Industrial Internet provides a whole new set of possibilities for us. For example, if the lights in a building are interconnected and also connected to motion sensors, the impact can be exponential. It is not just about having the lights turn off when the room is not in use, it's about harvesting data of facility usage to optimize other building systems such as HVAC and to further leverage efficiency.

From buildings and factories to cars, ships, locomotives, or entire cities, the insights produced by the Industrial Internet have the power to transform our world. And Miller Electric will be at the center of the convergence, capitalizing on the information and energy powering the possibility of a more connected future. As stewards of our business, it won't be enough to keep the status quo ... we must be aware of the changing world and innovate to keep pace. We owe it to our clients and we owe it to our future generations to keep looking forward.

Yes, all of this—and more—is the legacy we received through wise decisions made years ago. However, none of it will be accomplished without holding trust as our high standard and integral part of our culture. Healthy, productive, innovative relationships with our clients, vendor partners, and fellow employees all depend on it as the necessary ingredient enabling us to work together on the Miller Electric team. Trust provides room to move and grow, and allows all of us to perform at our very best.

Fishing in Palatka

With Emma - 2015

65th anniversary - 2015

Easter - 2016

Imparting skills to my great-grandson

Our new boat, a Marlow Cruiser

CPSIA information can be obtained
at www.ICGtesting.com
Printed in the USA
LVOW05*0202190118
563169LV00031B/132/P